THE PRINC

CRACKING THE REGENTS

SEQUENTIAL MATH II

THE·PRINCETON REVIEW

1998-99

CRACKING THE REGENTS

SEQUENTIAL MATH II

DOUGLAS FRENCH

1998–99 Edition
Random House, Inc.
New York, 1998
www.randomhouse.com

Princeton Review Publishing
2315 Broadway
New York, NY 10024
e-mail: info@review.com

Copyright 1998 by Princeton Review Publishing, L.L.C.

All rights reserved under International and Pan-American Copyright
Conventions.

Published in the United States by Random House, Inc., New York, and
simultaneously in Canada by Random House of Canada Limited, Toronto.

ISBN 0-375-75065-7

Editor: Lesly Atlas
Design & Production: Chris J. Thomas
Production Editor: Kristen Azzara

Manufactured in the United States of America on partially recycled paper.

9 8 7 6 5 4 3 2 1

1998–99 edition

ACKNOWLEDGMENTS

I'm especially grateful to Carl Hostnik and Frank Quinn, two of the best math professors of all time. (When working through really long calculus problems, Mr. Hostnik would routinely write on bulletin boards, walls, and posters when he ran out of space on the blackboard.) They taught me that a math teacher can be completely nuts and still be a great teacher.

Thanks to Melanie Sponholz and Evan Schnittman, the powers who got me involved with this project. My editors at The Princeton Review, including Lesly Atlas, Amy Zavatto, and Kristen Azzara, were a great help. Thanks also to the PageMakers, Chris J. Thomas and Robert McCormack, who made sure all the diagrams looked good, and to the expert reviewers, Sasha Alcott, Kenneth Butka, Blase Caruana, and Nancy Schneider, who made sure it all made sense.

Thanks to my family (especially Mom, who convinced me that I could sit down long enough to write a book), and to Agnes Dee, who is my favorite favorite.

And most of all, I'd like to thank all the math students I've tutored over the years. There's nothing better than getting a phone call from an ecstatic student who has just aced a final exam.

TABLE OF CONTENTS

PART I

INTRODUCTION

Welcome to *Cracking the Regents: Sequential Math II*! For too long now, students have not had a choice when it comes to studying for the various Regents tests. Well, we've finally come out with a book that explains the parts of the Sequential II Math exam in plain English and gives you useful hints for getting the grade you want.

HOW TO USE THIS BOOK

This book contains 10 real Sequential II Math Exams and the answers and explanations to every problem, complete with diagrams and strategies. Try as many tests as you can, so you can see how similar the tests are to each other. When you take each test, be sure to have the following:

- Scrap paper
- Graph paper
- Extra pencils
- Blue or black pen
- Ruler or straightedge
- Compass (the kind you draw with)
- Calculator (but not a graphing calculator, like a TI-82/83)
- This book

There's also a glossary and a brief chapter listing most of the stuff that shows up on the test year in and year out. If you can take every one of the tests in this book, you'll have a very good idea of what to expect on your exam.

THE TEST FORMAT

The Sequential II Regents Exam takes three hours, and it's composed of three parts:

	Number of Questions	Number to Do	Points Each	Total Points
Part One	35	30	2	60
Part Two	6	3	10	30
Part Three	2	1	10	10
			Total:	**100**

Grading Part One is easy: Two points if you're right, zero if you're not. Parts Two and Three, though, are a bit trickier. If possible, ask your teacher to grade them for you. (Chances are, you'll be taking a lot of these tests in class anyway, as the test nears.) Because this test differs very little from session to session, your scores on these practice exams will be an excellent indicator of the score you can expect on the Real Deal.

IT'S THE SAME OLD STUFF

The great thing about this test is that the same types of questions show up over and over again. There are very few surprises, and if you see something new, you can skip it. In fact, you may notice that some of the explanations in the book look a lot alike. That's because the questions are so alike!

DO'S AND DON'TS

The Board of Regents has some pretty stiff grading guidelines, which are published in a booklet that every test grader must read. Here are some rules of thumb you should know about.

Graphing

This test requires a lot of work with the x- and y-axes. Whenever you draw them, make sure they look like this:

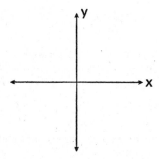

Also be sure to label each axis, so the grader knows that you know what you're talking about.

Graphing Parabolas

In Part Two, you'll usually have the option of graphing a parabola on the coordinate axes (in fact, of the ten tests in this book, nine ask you to graph a parabola).

Most of the time, they'll give you the values of x to plug into the equation. When they do, you'll always get seven coordinates—three on either side of the axis of symmetry. (For example, if the domain that you're given is $0 \le x \le 6$, you can bet your last buck that the axis of symmetry of the parabola is the line $x = 3$.)

Plug each of the seven x-coordinates into the equation, and you'll have seven points to plot on the coordinate axes. Once you're through plotting, the curve should be symmetrical (otherwise, you may have plotted a point incorrectly). When you connect them, don't just play connect-the-dots; you'll likely be penalized. Call upon whatever artistic skills you possess and make the line curved:

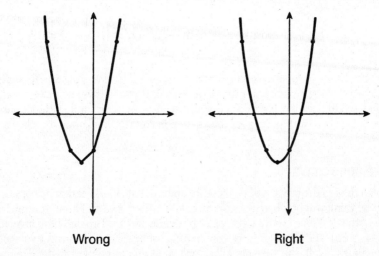

Wrong   Right

Calculators

Make sure your calculator has the following functions:

- Square root
- Trig functions (sine, cosine, tangent)
- Inverse trig functions ($\sin^{-1}$, etc.)

Be sure you know how to use each of these functions comfortably.

Do *not* bring a graphing calculator. With a TI-82/83 (or a comparable piece of machinery), you could whip through graphing problems in nothing flat. This makes the Regents people uneasy, since much of the test involves graphing on the coordinate axes.

Read Carefully and Follow Directions

Make sure you clearly indicate which questions you omit by writing "OMIT" on your answer sheet in the space provided. If they say "round to the nearest tenth," do it. If they want the positive root only, don't give them the negative root.

You're not under a lot of time pressure, so read each question completely and be sure to provide exactly what they ask for—no more, no less. You could lose full credit if you don't.

Example:

5. What is the positive root of the equation $x^2 - 3x - 10 = 0$?

There are two roots to this equation: 5 and –2. The correct answer, though, is 5 because they ask only for the positive root. If your answer was $x = \{5, -2\}$, you wouldn't get any credit (0 points!) because you didn't follow directions. So, before you start looking for the answer, make sure you know what they're asking you for.

THE GOOD NEWS

Even though the graders of this test are notorious sticklers for detail, fear not. If you think you bombed an entire problem in Part Two or Three, you might be better off than you thought. Most problems in Part Two or Three involve many steps, and the later work you do usually depends on an answer you got previously. Well, the Regents folks don't want you to be punished twice for one mistake. Even if the numbers you used on a question are wrong, you could still get full credit if you did the math correctly.

Here's a sample problem from Part Two:

39. In the accompanying diagram of trapezoid *ABCD*, $\overline{CD} \perp \overline{AD}$, *BC* = 9, *AD* = 15, and m∠*A* = 35.

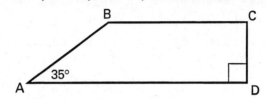

Find, to the *nearest tenth*, the

a area of *ABCD*   [5]

b perimeter of *ABCD*   [5]

To find the area of the trapezoid, you need to find the length of $\overline{CD}$ using trigonometry. (The answer is 4.2.) Suppose you made a mathematical mistake and got 8.4 instead, then you used this to figure out the answer to part *b*. You would lose the first five points, but the grader might give you the five points in part *b* if you did the math right.

HELPFUL HINTS

The people who write the Sequential II Math exam like to think of themselves as The Big Bad Regents Board. Well, we're here to tell you that this test can be really easy—if you take the time to prepare. Here are some tips to help you perform at your best.

Relax!

Unlike most other timed exams, including the SAT and PSAT, you probably won't find yourself having to rush. They give you three hours to finish the exam, and most students don't need more than two. This lack of time pressure gives you the chance to check your work.

Always Double-Check

Errors happen, so check everything twice. (After all, you've got the time.) A typical question might involve finding the roots of an equation, like so:

$$x^2 - 4x - 21 = 0$$
$$(x - 7)(x + 3) = 0$$
$$x = \{7, -3\}$$

When you've finished with the algebra, take the time to check your answers by plugging them back into the equation:

$$(7)^2 - 4(7) - 21 = 0 \qquad (-3)^2 - 4(-3) - 21 = 0$$
$$49 - 28 - 21 = 0 \qquad 9 + 12 - 21 = 0$$
$$0 = 0 \qquad 0 = 0$$

With just a few seconds of extra work, you can be sure you did the work correctly.

Backsolving

This test-taking technique will help you to check your work. Sometimes you can find the right answer by plugging the answer choices you're given back into the question. Here's a sample problem:

17. The roots of the equation $2x^2 - 7x - 4 = 0$ are

(1) $\dfrac{7 \pm \sqrt{17}}{4}$

(2) $-\dfrac{1}{2}, 4$

(3) $\dfrac{1}{2}, -4$

(4) $\dfrac{-7 \pm \sqrt{17}}{2}$

Rather than factor this right away, try a little backsolving. One of the answer choices has to work, so try one of the easier numbers first. (Don't bother with (1) and (4) yet—they're too hard!)

Since 4 is the only root given that is an integer, try it first (use your calculator if you feel the urge):

$$2(4)^2 - 7(4) - 4 = 0$$
$$2(16) - 28 - 4 = 0$$
$$32 - 32 = 0$$

Bingo! Since answer choice (2) is the only choice that has 4 in it, you know that (2) is the correct answer. And all you did was a little simple arithmetic!

Note: This technique works ONLY in the multiple-choice section of Part One. Parts Two and Three require you to show all your work, so you can't take short cuts.

PROCESS OF ELIMINATION (POE)

The questions on the second portion of Part One have four answer choices. On these questions, you can sometimes choose the right answer just by eliminating the other three possibilities.

18. The coordinates of two points are $X(2, 5)$ and $Y(6, 5)$. Which of the following equations represents the locus of points that is the perpendicular bisector of $\overline{XY}$?

    (1) $x = 4$ (3) $x = 8$
    (3) $y = 4$ (4) $y = 8$

If the answer to this one eludes you for the moment, don't panic. Graph the points first:

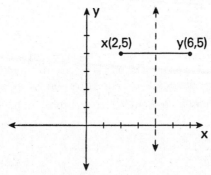

Your first instinct might be to find the midpoint of the segment. Don't bother; there's a way to solve this without doing any mathematical calculation.

From the diagram, you can see that segment $\overline{XY}$ is horizontal; the perpendicular bisector must be vertical. Therefore, you can eliminate answer choices (2) and (4), which are also horizontal. If you guess at this point, you have a 50-50 chance. Those odds aren't bad.

If you look further, you'll realize that the perpendicular bisector has to be between the two points. Since answer choice (4) is too far to the right, you can eliminate it. The only answer left is choice (1).

This is why you should take a long look at every question in the multiple-choice section of Part One. If you don't see a method right away, you still have a chance using POE.

WHEN TO OMIT

Whatever you do, don't be too quick to skip a question. If you're not quite sure how to answer something, come back to it later. Don't rush to omit a question unless you have absolutely no clue.

LOGIC

Geometric and logical proofs are the reason why this is the only Regents math test with a Part Three.

It would be very hard to escape this test without having to construct a proof, so study your theorems. As you'll see from the proofs in this book, there are only a couple dozen or so theorems, postulates, and definitions to know. As far as logical proofs go, there are seven things to know (see the Stuff You Should Know chapter). Becoming familiar with them shouldn't be too difficult—so do it!

GEOMETRIC PROOFS

When you plan a geometric proof, try to visualize how you're going to end it before you put pencil to paper. This will keep you from rambling on, spouting a bunch of irrelevant theorems, and hoping for some degree of partial credit.

Here's a sample proof question that would probably appear in Part Three:

Given: $\overline{DB} \perp \overline{AC}$; $\angle 1 \cong \angle 2$

Prove: $\overline{AB} \cong \overline{BC}$

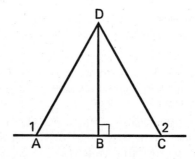

Analyze the given statements. $\overline{AB}$ and $\overline{BC}$ are corresponding parts of $\triangle ABD$ and $\triangle CBD$, so how can you prove that the two triangles are congruent? You can prove that $\angle DAB$ is congruent to $\angle DCB$, because they're supplemental to $\angle 1$ and $\angle 2$, respectively. Since $\overline{DB}$ is perpendicular to $\overline{AC}$, angles

$\angle BA$ and DBC are right angles, which are always congruent. Further, the triangles share a common side, $\overline{DB}$.

From this information, you can plan to prove that $\triangle ABD$ and $\triangle CBD$ are congruent using Angle-Angle-Side, then use CPCTC. Write that on the page first, and don't worry about numbering. You can number everything when you're finished.

Statements	Reasons
$\triangle DBA \cong \triangle DBA$	$AAS \cong AAS$
$\overline{AB} \cong \overline{BC}$	CPCTC.

Now that you have a clear objective, you can fill in the blanks as you go along. This minimizes the chance that you'll include a lot of stuff you don't need. Here's the final proof:

Statements	Reasons
1. $\angle 1 \cong \angle 2$	1. Given
2. $\angle 1$ and $\angle DAB$ are supplemental; $\angle 2$ and $\angle DCB$ are supplemental	2. Definition of supplemental angles
3. $\angle DAB \cong \angle DCB$	3. Angles that are supplemental to congruent angles are congruent.
4. $\overline{DB} \bot \overline{AC}$	4. Given
5. $\angle DBA$ and $\angle DBC$ are right angles	5. Definition of perpendicular lines
6. $\angle DBA \cong \angle DBC$	6. All right angles are congruent
7. $\overline{DB} \cong \overline{DB}$	7. Reflexive property of equality
8. $\triangle DBA \cong \triangle DBC$	8. AAS $\cong$ AAS
9. $\overline{AB} \cong \overline{BC}$	9. CPCTC

Look at Reason 5. Some teachers might want you to write that "perpendicular lines intersect to form four right angles," while others are just as happy with "definition of perpendicular lines." When you list your reasons, phrase them in the same way that your instructor has taught you all year. This is no time to get creative.

THE DAY BEFORE THE TEST

Don't try to cram a lot of information into your brain the night before you take the test. After all, what extra bit of magic can you learn that you didn't learn all year? Look over a few formulas, make sure you have everything you need to take to the test site, and get a good night's sleep. Cramming the night before just makes you flaky and burns you out the next day.

THE DAY OF THE TEST

There's a big difference between "awake" and "alert." Get up early and have something to eat. Your body wakes up when it has to digest food, and your breakfast will serve as an energy source throughout the morning (or afternoon, if you're a late sleeper).

Bring several layers of clothing to the test. You don't know if it will be too warm or too cold at the test site, so be prepared. You'll do much better if you're comfortable.

Get to the test site early. While you're waiting, take out this book and look over a few problems you did well on. (*Don't* try anything new.) As you familiarize yourself with the material once again, you'll get your mind ready to do math for three hours.

Take your time, be thorough, and crush this test. It'll be over before you know it.

Good luck!

STUFF YOU SHOULD KNOW

This chapter lists the formulas and other calculations you should be familiar with for the Sequential II Math exam, as well as a few handy ways to remember them. If a complete explanation of something doesn't appear here, it probably appears in full within the explanations of one of the tests. You can also check the Glossary.

Some teachers let their students bring a sheet of formulas in with them to the exam. Ask your teacher if he or she will let you do so. Otherwise, *memorize*.

ALGEBRA

You should have a good sense of factoring quadratics in order to determine their roots. For example:

$$x^2 - 4x - 12 = 0$$
$$(x - 6)(x + 2) = 0$$
$$x = \{6, -2\}$$

If you can't factor a quadratic equation, you can always use the Quadratic Formula. Given the equation $ax^2 + bx + c = 0$, you can find the roots using this:

$$x = \frac{-b \pm \sqrt{b^2 - 4ac}}{2a}$$

It's not mandatory, but it might be helpful in some places to remember that the sum of the roots of a quadratic in standard form is $-\frac{b}{a}$, and the product is $\frac{c}{a}$.

AREAS

Know these area formulas:

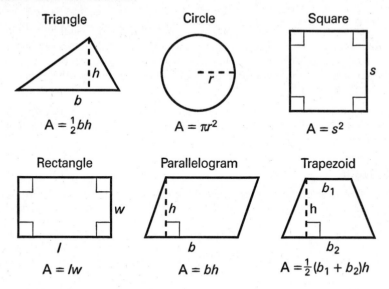

Triangle	Circle	Square
$A = \frac{1}{2}bh$	$A = \pi r^2$	$A = s^2$

Rectangle	Parallelogram	Trapezoid
$A = lw$	$A = bh$	$A = \frac{1}{2}(b_1 + b_2)h$

CONSTRUCTIONS

The following constructions have appeared on the Regents Exam:

- Perpendicular bisector of a segment (page 108)
- Angle bisector (page 69)
- Duplicate angle (page 32)
- Parallel line through a point
- Perpendicular line though a point (page 304)

Each of these is explained fully, with diagrams, in the test explanations.

EQUATIONS

The most basic graphic equation you should be able to recognize is the standard form of a line, in which m represents the slope of the line and b is the y-intercept:

$$y = mx + b$$

We use this formula in this book (instead of the form $y - y_1 = m(x - x_1)$, which you might have used throughout your school year) because the exam always uses standard form. The same is true for the equation of a parabola; the standard form looks like this:

$$y = ax^2 + bx + c$$

This format is used on the exam instead of the form $y - k = a(x - h)^2$ because it's easier to factor. You won't have to worry about foci or the directrix.

The formula for a circle, however, does use the (h, k) format:

$$(x - h)^2 + (y - k)^2 = r^2$$

in which (h, k) is the center of the circle and r is the radius.

FORMULAS

Here are the three big formulas you need to know, given the points (x_1, y_1) and (x_2, y_2):

The distance between them is:

$$d = \sqrt{(x_2 - x_1)^2 + (y_2 - y_1)^2}$$

Think of the distance formula as a manifestation of the Pythagorean Theorem. The distance between the points is the hypotenuse of a right triangle:

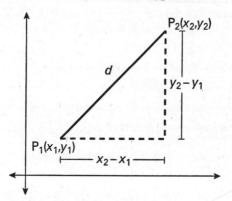

The formula for the slope between the points is:

$$m = \frac{y_2 - y_1}{x_2 - x_1}$$

The slope of a line represents the $\frac{rise}{run}$. If you rer

the y's," you'll remember to put the difference of the ence of the x's on the bottom.

The midpoint formula is:

$$(\overline{x},\overline{y}) = \left(\frac{x_1 + x_2}{2} , \frac{y_1 + y_2}{2} \right)$$

This one's easy to remember, because it makes sense that the point equidistant from two others is just the average of their x- and y-coordinates.

GEOMETRIC PROOFS

Know how to use each of these triangle congruence theorems:

- Angle-Angle-Side (AAS)
- Angle-Side-Angle (ASA)
- Side-Angle-Side (SAS)
- Side-Side-Side (SSS)
- Hypotenuse-Leg (HL)—for right angles only

You should also be familiar with the Angle-Angle theorem for similarity, as well as the theorem that "corresponding parts of congruent triangles are congruent" (CPCTC). As far as the definitions and properties you need to know, consult your textbook and the explained proofs of this book. You'll see that a select few properties show up a lot, such as:

- Reflexive Property of Equality
- Addition and Subtraction Properties of Equality
- Definition of midpoint and bisector
- Definition of perpendicular lines and right angles
- Supplements and complements of angles

Memorize the characteristics of each of the following:

- Squares
- Rectangles
- Rhombuses (or Rhombi)
- Parallelograms
- Isosceles triangles
- Isosceles trapezoids

The Introduction contains some helpful hints for writing geometric proofs.

GRAPHING

Obviously, knowing how to plot a point (x, y) on the coordinate axes is a must. You'll use this knowledge to plot parabolas and the vertices of various polygons.

Remember that the formula for the axis of symmetry of the parabola $y = ax^2 + bx + c$ is:

$$x = -\frac{b}{2a}$$

You should also recognize the various formulas for transformations, including:

- Translations, under which a point (x, y) "slides" to the point $(x + h, y + k)$
- Dilations, under which a point (x, y) "expands" to the point (kx, ky)
- Reflections in the following:
  1. $r_{x-axis}(x, y) \rightarrow (x, -y)$
  2. $r_{y-axis}(x, y) \rightarrow (-x, y)$
  3. $r_{(0,0)}(x, y) \rightarrow (-x, -y)$
  4. $r_{y=x}(x, y) \rightarrow (y, x)$

As you'll see, there's been nothing on any of these 10 tests involving rotations. That stuff usually appears on the Sequential III exam.

LOGIC

Know these rules (and their translations), and you'll be fine:

- **Law of Contrapositive Inference:** if $A \rightarrow B$, then $\sim B \rightarrow \sim A$ (if A leads to B, then without B there is no A)
- **The Chain Rule:** if $A \rightarrow B$ and $B \rightarrow C$, then $A \rightarrow C$
- **The Law of Detachment (*Modus Ponens*):** $[(A \rightarrow B) \wedge A] \rightarrow B$ (if A leads to B, and A is true, then B is true)
- **The Law of *Modus Tollens*:** $[(A \rightarrow B) \wedge \sim B] \rightarrow \sim A$ (if A leads to B, and B is not true, then A is not true; this is related to the contrapositive)
- **De Morgan's Laws:** $\sim(A \wedge B) \rightarrow \sim A \vee \sim B$ (if A and B are not both true, then either A is false or B is false or both); $\sim(A \vee B) \rightarrow \sim A \wedge \sim B$ (if neither A nor B is true, then both A and B must be false)
- **Law of Disjunctive Inference:** $[(A \vee B) \wedge \sim A] \rightarrow B$ (if either A or B is true and A is false, then B is true)
- **Law of Double Negation:** $\sim(\sim A) \rightarrow A$

See the logic proofs in this book for more practice.

PARALLEL LINES

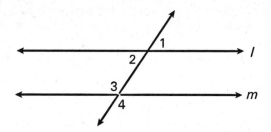

Given two parallel lines cut by a transversal, you should know that:

- Alternate interior angles are congruent ($\angle 2 \cong \angle 4$)
- Corresponding angles are congruent ($\angle 1 \cong \angle 4$)
- Interior angles on the same side of a transversal are supplementary ($m\angle 2 + m\angle 3 = 180$)

PROBABILITY AND PERMUTATIONS

Probability shows up on a few questions on each exam. Remember that the probability that something will happen is denoted by the number of favorable outcomes divided by the number of possible outcomes. For example, the probability that you'll get an even number when you roll an ordinary, six-sided die is

$\dfrac{3}{6}$, because there are six possible rolls and three even numbers (2, 4, and 6).

When something is certain to happen, the probability that it will happen is one. When something can't happen, the probability is zero.

Combinations show up a lot more than permutations, because order rarely matters on Sequential II test questions. For the record, though, here are the formulas:

$$_nP_r = \frac{n!}{r!} \qquad\qquad _nC_r = \frac{n!}{r!(n-r)!}$$

Problems involving combinations of letters in a word occur rather frequently. In a word with n letters, in which one letter appears p times and another appears q times (given that p and q are greater than 1), the formula is:

$$\frac{n!}{p!\,q!}$$

TRIANGLES

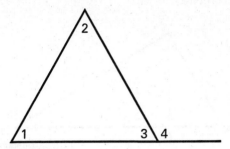

The Rule of 180 (m ∠1 + m ∠2 + m ∠3 = 180) is tested often; also don't forget that the measure of an exterior angle equals the sum of the measures of the two non-adjacent interior angles (m ∠4 = m ∠1 + m ∠2).

Also, remember that the length of a side of a triangle must be greater than the difference between the lengths of the two other sides and less than their sum:

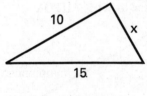

5 < x < 25

As far as right triangles are concerned, the Pythagorean Theorem comes up a lot. It's also helpful to know the relationships between the sides of 30:60:90 and 45:45:90 triangles:

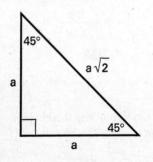

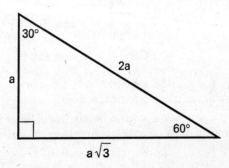

When an altitude in a right triangle is drawn, three similar triangles are created with proportional lengths:

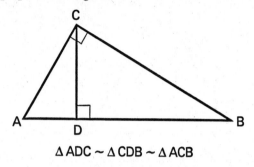

$$\triangle ADC \sim \triangle CDB \sim \triangle ACB$$

You'll usually get a question about the largest side or angle of a triangle. In any triangle, the biggest side is always opposite the biggest angle, and the smallest side is opposite the smallest angle.

TRIGONOMETRY

On this exam, trig is limited to SOHCAHTOA:

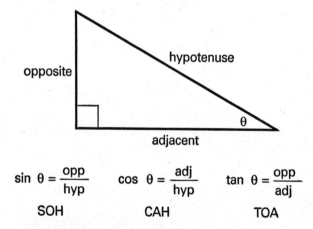

$$\sin \theta = \frac{opp}{hyp} \qquad \cos \theta = \frac{adj}{hyp} \qquad \tan \theta = \frac{opp}{adj}$$

SOH   CAH   TOA

You should also know how to use your calculator to find values and inverse values of trigonometric functions.

If you have a good understanding of the terms and formulas in this chapter, you are well on your way to getting the score you want on the Sequential II test.

PART II

EXAMINATIONS
AND
EXPLANATIONS

EXAMINATION: JANUARY 1994

Part I

Answer 30 questions from this part. Each correct answer will receive 2 credits. No partial credit will be allowed. Write your answers in the spaces provided on the separate answer sheet. Where applicable, answers may be left in terms of π or in radical form. [60]

1 The lengths of the sides of a triangle are 4, 6, and 7. If the length of the longest side of a similar triangle is 21, find the perimeter of the larger triangle.

2 The measure of an exterior angle of a triangle is 120°, and the measure of one interior angle of the triangle is 50°. Find the number of degrees in the measure of the largest angle of the triangle.

3 Using the accompanying table, solve for y if $a \heartsuit y = c \heartsuit d$.

$\heartsuit$	a	b	c	d
a	b	c	d	a
b	c	d	a	b
c	d	a	b	c
d	a	b	c	d

4   If sin A = 0.3642, find the measure of ∠A to the *nearest degree.*

5   In the accompanying diagram ΔABC, D is the midpoint of $\overline{AB}$ and E is the midpoint of $\overline{BC}$. If DE = 5 and AC = 2x − 20, find x.

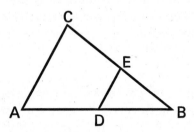

6   If one root of the equation $x^2 + 5x + c = 0$ is −2, find the value of c.

7   In the accompanying diagram, ΔIHJ ~ ΔLKJ. If IH = 5, HJ = 2, and LK = 7, find KJ.

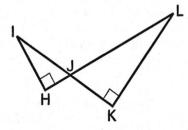

8   Express, in radical form, the distance between the points whose coordinates are (2,4) and (−2,5).

9 Solve for x: $\dfrac{2}{x} + \dfrac{1}{7} = \dfrac{4}{x}$, $x \neq 0$

10 Solve for all values of x: $\dfrac{6}{x-1} = \dfrac{x}{2}$, $x \neq 1$

11 In the accompanying diagram of right triangle ABC, m $\angle C = 90$, m $\angle A = 45$, and $AC = 1$. Find, in radical form, the length of $\overline{AB}$.

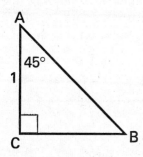

12 What is the total number of ways a committee of three can be chosen from a group of five people?

13 Express in lowest terms: $\dfrac{x^2 - 9}{x^2 + 3x}$, $x \neq 0, -3$

Directions (14-34): For each question chosen, write the numeral preceding the word or expression that best completes the statement or answers the question.

14 The y-intercept of the line whose equation is $y = 3x - 2$ is

(1) –2

(2) 2

(3) 3

(4) $\dfrac{1}{3}$

15 In the accompanying diagram, line l is parallel to line m, and lines s and t are transversals that intersect at a point on line m.

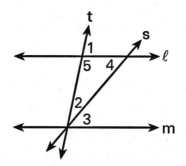

Which statement must be true?

(1) $m\angle 1 = m\angle 4$

(2) $m\angle 4 = m\angle 2$

(3) $m\angle 1 = m\angle 2 + m\angle 3$

(4) $m\angle 5 = m\angle 2 + m\angle 3$

16 Given: line l and point P not on l. According to the Euclidean parallel postulate, how many lines pass through P that are parallel to l?

(1) 1
(2) 2
(3) an infinite number
(4) 0

17 Which is an equation of the locus of points whose coordinates are three less than twice their abscissas?

(1) $y = 2x + 3$ (3) $x = 2y + 3$
(2) $y = 2x - 3$ (4) $x = 2y - 3$

18 A translation moves point $A(4,-2)$ onto point $A'(0,2)$. What is the image of (x,y) under this translation?

(1) $(x + 2,y)$ (3) $(x - 2,y + 2)$
(2) $(x - 4,y + 4)$ (4) $(x + 4,y + 2)$

19 Which statement would never be used to prove that a figure is a rhombus?

(1) The figure is a quadrilateral.
(2) The figure has a pair of equal adjacent sides.
(3) The figure is a parallelogram.
(4) The figure is a rectangle.

20 The diagonals of a rhombus are 10 centimeters and 24 centimeters. A side of the rhombus measures

(1) 10 cm   (3) 24 cm
(2) 13 cm   (4) 26 cm

21 Given: points $A(1,2)$, $B(4,5)$, and $C(6,7)$. Which statement is true?

(1) AB is equal to BC.
(2) AB is perpendicular to BC.
(3) Points A, B, and C are collinear.
(4) The slope of $\overline{AC}$ is –1.

22 In $\triangle QRS$, $m\angle Q = x$, $m\angle R = 8x - 40$, and $m\angle S = 2x$. Which type of triangle is $\triangle QRS$?

(1) isosceles   (3) acute
(2) right   (4) obtuse

23 Which statement is logically equivalent to $\sim(q \wedge \sim s)$?

(1) $\sim q \wedge s$ (3) $\sim q \vee s$
(2) $q \wedge \sim s$ (4) $q \vee \sim s$

24 The coordinates of the image of $P(3,-4)$ under a reflection in the x-axis are

(1) $(3,-4)$ (3) $(3,4)$
(2) $(-3,4)$ (4) $(-3,-4)$

25 Which is an equation of a line that is perpendicular to the line whose equation is

$$y = \frac{1}{3}x - 2?$$

(1) $y = -3x + 2$ (3) $y = -\frac{1}{3}x = 2$

(2) $y = 3x + 2$ (4) $y = \frac{1}{3}x + 2$

26 If the lengths of two sides of a triangle are 4 and 7, the length of the third side can *not* be

(1) 11   (3) 5
(2) 7   (4) 4

27 Which is an equation of the circle whose center is $(0,4)$ and whose radius is 3?

(1) $x^2 + (y - 4)^2 = 3$
(2) $x^2 + (y - 4)^2 = 9$
(3) $(x - 4)^2 + (y - 3)^2 = 9$
(4) $(x - 4)^2 + y^2 = 9$

28 A set contains four distinct quadrilaterals: a parallelogram, a rectangle, a rhombus, and a square. If one quadrilateral is selected from the set at random, what is the probability that the diagonals of that quadrilateral bisect each other?

(1) 1   (3) $\dfrac{2}{4}$

(2) $\dfrac{1}{4}$ (4) $\dfrac{3}{4}$

29 Which transformation moves (x,y) to $(5x,5y)$?

(1) reflection   (3) translation
(2) rotation   (4) dilation

30 In the accompanying diagram, $\overline{MN} \perp \overline{NP}$, $\overline{QP} \perp \overline{PN}$, O is the midpoint of $\overline{NP}$, and $\overline{MN} \cong \overline{QP}$.

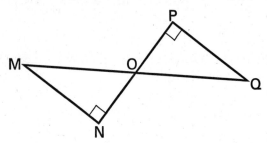

Which reason would be least likely to be used to prove $\triangle MNO \cong \triangle QPO$?

(1) $HL \cong HL$
(2) $AAS \cong AAS$
(3) $SAS \cong SAS$
(4) $ASA \cong ASA$

31 Which is an equation of the axis of symmetry of the graph of the equation $y = 2x^2 - 5x + 3$?

(1) $x = -\dfrac{5}{2}$ (3) $x = -\dfrac{5}{4}$

(2) $x = \dfrac{5}{2}$ (4) $x = \dfrac{5}{4}$

32 In the accompanying diagram, the legs of right triangle *ABC* are 5 and 12 and the hypotenuse is 13.

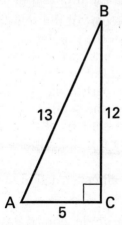

What is the value of cos *A*?

(1) $\dfrac{12}{13}$ (3) $\dfrac{5}{13}$

(2) $\dfrac{13}{5}$ (4) $\dfrac{12}{5}$

33 Given the true statements:

$\sim a \lor \sim b$

b

$c \to a$

Which statement is also true?

(1) c (3) $\sim c$
(2) $\sim b$ (4) a

34 Which statement is logically equivalent to the statement: "If you are not part of the solution, then you are part of the problem"?

(1) If you are part of the solution, then you are not part of the problem.
(2) If you are not part of the problem, then you are part of the solution.
(3) If you are part of the problem, then you are not part of the solution.
(4) If you are not part of the problem, then you are not part of the solution.

Directions (35): Show all construction lines.

35 Construct an angle congruent to angle *B* of hexagon *ABCDEF*, using point *W* as the vertex.

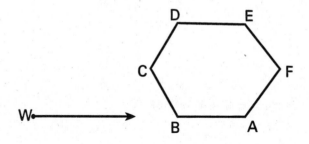

Part II

Answer *three* questions from this part. Clearly indicate the necessary steps, including appropriate formula substitutions, diagrams, graphs, charts, etc. Calculations that may be obtained by mental arithmetic or the calculator do not need to be shown. [40]

36 *a* On graph paper, draw the graph of the equation $y = x^2 - 4x + 3$, including all values of x in the interval $-1 \le x \le 5$. [4]

b On the same set of axes, draw the graph of the image of the graph drawn in part *a* after the translation which moves (x,y) to $(x + 3, y + 2)$, and label this graph *b*. [3]

c On the same set of axes, draw the graph of the image of the graph drawn in part *b* after a reflection in the *x*-axis, and label this graph *c*. [3]

37 *a* For all values of *x* for which these expressions are defined, express the product in simplest form:

$$\frac{4x - 24}{x^2 - 36} \bullet \frac{x^2 + 4x - 12}{x^2 + x - 6} \quad \text{[4]}$$

b Solve the following system of equations:

$$y = x + 5$$
$$x^2 + y^2 = 97 \quad \text{[6]}$$

38 In the accompanying diagram of $\triangle ABC$, D is a point on $\overline{AB}$ and E is a point on $\overline{BC}$ such that $\overline{DE}$ is parallel to $\overline{AC}$. If DB is 2 less than AD, AC is 5 more than AD, and $DE = 4$, find the length of $\overline{AD}$ to the *nearest tenth*. [5,5]

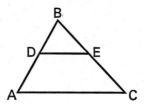

39 As shown in the accompanying diagram, a ship is headed directly toward a coastline formed by a vertical cliff $\overline{BC}$, 70 meters high. At point A, the angle of elevation from the ship to B, the top of the cliff, is 23°. A few minutes later at point D, the angle of elevation increased to 30°.

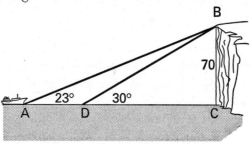

a To the nearest meter, find:

    (1) *DC* [3]

    (2) *AC* [3]

    (3) *AB* [3]

b To the nearest meter, what is the distance between the ship's position at the two sightings? [1]

40 Given: If Ronnie does not waste time in class,
               then she does well in Course II.

               If Ronnie is absent from class, then her
               grades will go down.

               Either Ronnie does not waste time in class
               or Ronnie is absent from class.

               Ronnie's grades do not go down.

       Let T represent: "Ronnie wastes time in class."

       Let A represent: "Ronnie is absent from class."

       Let S represent: "Ronnie's grades go down."

       Let B represent: "Ronnie does well in Course II."

       Prove: Ronnie does well in Course II.   [10]

Part III

Answer *one* question from this part. Clearly indicate the necessary steps, including appropriate formula substitutions, diagrams, graphs, charts, etc. Calculations that may be obtained by mental arithmetic or the calculator do not need to be shown. [10]

41 Given: isosceles triangle ABC, $\overline{BA} \cong \overline{BC}$, $\overline{AE} \perp \overline{BC}$, and $\overline{BD} \perp \overline{AC}$.

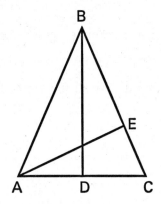

Prove: $\dfrac{AC}{BA} = \dfrac{AE}{BD}$ [10]

42 Quadrilateral $MATH$ has vertices $M(-1,4)$, $A(4,7)$, and $H(2,-1)$. Prove that $MATH$ is a square. [10]

ANSWER KEY

Part I

1. 51

2. 70

3. b

4. 21°

5. 15

6. 6

7. $\dfrac{14}{5}$, or 2.8

8. $\sqrt{17}$

9. 14

10. 4, −3

11. $\sqrt{2}$

12. 10

13. $\dfrac{x-3}{x}$

14. 1

15. 3

16. 1

17. 2

18. 2

19. 4

20. 2

21. 3

22. 4

23. 3

24. 3

25. 1

26. 1

27. 2

28. 1

29. 4

30. 1

31. 4

32. 3

33. 3

34. 2

35. construction

EXPLANATIONS:
JANUARY 1994

1. 51

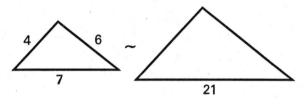

From the diagram, you can see that the largest side of the smaller triangle is 7, and its counterpart of the larger triangle is 21 units long. Now you know that each side of the larger triangle is three times as big as its corresponding side of the smaller triangle. The perimeter of the larger triangle must also be three times the perimeter of the smaller triangle.

The perimeter of the small triangle is 4 + 6 + 7, or 17. Therefore, the perimeter of the larger triangle must be 17 × 3, or 51.

2. 70

Make a drawing of the situation first:

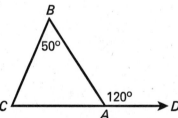

Exterior angle *DAB* measures 120°, so interior angle *BAC* measures 60°. Since the measure of ∠*B* is 50°, you can find the measure of ∠*C* using the Rule of 180:

$$m\angle BAC + m\angle B + m\angle C = 180$$
$$60 + 50 + m\angle C = 180$$
$$m\angle C = 70.$$

∠*C* is the largest, and it measures 70°

3. b

To find the value of y in the equation, you first must find the value of $(c \heartsuit d)$. Find c along the far left column, then follow the row along until you reach Column d. Row c intersects Column d at Point c.

Therefore, $(a \heartsuit y) = c$. Follow Column a along until you find c; there's a c in Column b. Since $(a \heartsuit b) = c$, it must be true that $y = b$.

4. $21°$

Your calculator makes this a snap. Just enter 0.3642 and then press the inverse-sine button (which usually looks like "sin⁻¹"). You'll get 21.36. (If you didn't get this, make sure your calculator is in "degree" mode.) Since they want you to round off the answer to the nearest degree, your answer is $21°$.

5. **15**

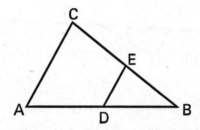

Here's a test of a seldom-tested geometry theorem involving triangles. Since E is the midpoint of $\overline{BC}$ and D is the midpoint of $\overline{AB}$, segment DE is a line segment that joins the midpoints of a triangle. Therefore, $\overline{DE}$ is parallel to $\overline{AC}$ and is also half its length. Set up the equation:

$$DE = \frac{1}{2} AC$$
$$5 = \frac{1}{2}(2x - 20)$$
$$5 = x - 10$$
$$x = 15$$

If you didn't remember that theorem, you could have solved it using similar triangles. For example, BE is half as long as BC and BD is half as long as BA. Therefore, DE is half as long as AC.

6. 6

Since -2 is a root of the equation, you can plug it into the equation to determine the value of c, like so:

$$(-2)^2 + 5(-2) + c = 0$$
$$4 + (-10) + c = 0$$
$$-6 + c = 0$$
$$c = 6$$

7. $\dfrac{14}{5}$, or 2.8

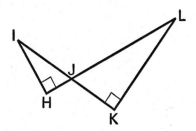

Wow. Three of the first seven problems involve similar triangles!

At least they had the decency to tell you the two triangles are similar. They could have made you figure it out using Angle-Angle. (Remember that $\angle IJH$ and $\angle LJK$ are congruent because they're vertical angles.)

Just set up the proportion between corresponding sides. The key is recognizing which sides correspond with which:

$$\frac{IH}{LK} = \frac{HJ}{JK}$$
$$\frac{5}{7} = \frac{2}{KJ}$$

Now cross-multiply:

$$5(KJ) = 7 \times 2$$
$$KJ = \frac{14}{5}$$

8. $\sqrt{17}$

To find the distance between two points, use the distance formula:

$$d = \sqrt{(x_2 - x_1)^2 + (y_2 - y_1)^2}$$
$$= \sqrt{(-2 - 2)^2 + (5 - 4)^2}$$
$$= \sqrt{(-4)^2 + (1)^2}$$
$$= \sqrt{16 + 1}$$
$$= \sqrt{17}$$

9. 14

You can't do anything to this equation yet, because the fractions have different denominators. The easiest way to solve it is to find a common denominator by multiplying the different denominators together: $7 \times x = 7x$.

Multiply every term in the equation by $7x$:

$$(7x)\frac{2}{x} + (7x)\frac{1}{7} = (7x)\frac{4}{x}$$

Now you can cancel out all the denominators and solve for x:

$$(7x)\frac{2}{x} + (7x)\frac{1}{7} = (7x)\frac{4}{x}$$
$$14 + x = 28$$
$$x = 14$$

Once you're finished, plug 14 back into the equation to make sure it works.

10. $x = \{4, -3\}$

When two fractions are equal to each other, you can cross-multiply them:

$$\frac{6}{x-1} = \frac{x}{2}$$
$$x(x - 1) = 6 \times 2$$
$$x^2 - x = 12$$
$$x^2 - x - 12 = 0$$

Now you have to factor the equation to find its roots:

$$(x - 4)(x + 3) = 0$$
$$x = \{4, -3\}$$

Once you're finished, plug 4 and –3 into the equation to make sure they both work.

11. $\sqrt{2}$

If you know your "special" triangles, this one is a piece of cake. Since $\angle C$ is a right angle and m $\angle A$ = 45, the Rule of 180 tells you that m $\angle B$ = 45 (90 + 45 + 45 = 180). Triangle ABC is a 45:45:90 triangle, so the lengths of each leg are the same, and the length of the hypotenuse equals the length of a leg times $\sqrt{2}$. Since $AC = 1$, the length of the hypotenuse, $\overline{AB}$, is $\sqrt{2}$.

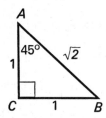

If you never studied 45:45:90 triangles, you should figure out the measure of $\angle B$, recognize that $\triangle ABC$ is isosceles, and find that BC also equals 1. Then use the Pythagorean Theorem.

12. 10

The order in which you choose the three people for the committee doesn't matter, so use the combinations formula:

$$_nC_r = \frac{n!}{r!\,(n - r)!}$$

in which n represents the number from which you have to choose and r represents the number of people you're choosing. You have five people and you're choosing three, so $n = 5$ and $r = 3$:

$$_5C_3 = \frac{5!}{3!\,2!} = \frac{5 \times 4 \times 3 \times 2 \times 1}{3 \times 2 \times 1 \times (2 \times 1)} = 5 \times 2 = 10$$

13. $\dfrac{x-3}{x}$

Factor the numerator and the denominator of the fraction like this:
$$x^2 - 9 = (x - 3)(x + 3)$$
$$x^2 + 3x = x(x + 3)$$

Now the fraction looks like this:

$$\frac{(x - 3)(x + 3)}{x(x + 3)}$$

Once you cancel the $(x + 3)$ terms, you're left with $\dfrac{x-3}{x}$.

Multiple Choice

14. (1)

When a line is expressed in the standard form $y = mx + b$ (as this one is), b represents the value of the y-intercept. Therefore, the y-intercept of $y = 3x - 2$ is -2.

15. (3)

To answer this problem, try to visualize the diagram without line s, like this:

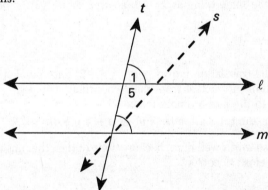

Lines l and m are parallel, and line t is a transversal. Therefore, $\angle 1$ and the angle that is the sum of $\angle 2$ and $\angle 3$ are corresponding angles. Corresponding angles formed by parallel lines have equal measure, so m $\angle 1$ = m $\angle 2$ + m $\angle 3$.

16. (1)

This is basic geometry. Don't let the reference to the "Euclidean parallel postulate" freak you out.

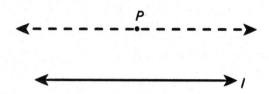

Through a point *not* on a line, there exists one (and only one) line that goes through the point and is parallel to the line. It's the same theorem, regardless of any fancy names that are attached to it.

17. (2)

The key thing to remember here is that "abscissa" is another word for the *x*-coordinate, and "ordinate" means the *y*-coordinate. (Memorize this. It doesn't come up often, but it's an easy 2 points if you remember it.)

Now, look at the question again; it mentions "twice their abscissas," which is another way of saying "two times *x*", or $2x$. Eliminate answer choices (3) and (4). Since "*three less than* twice their abscissas" is the same thing as $2x - 3$, answer choice (2) is the right answer.

18. (2)

Since the translation moves point $A(4, -2)$ onto point $A'(0, 2)$, the translation subtracts 4 from the *x*-coordinate (since $4 - 4 = 0$) and adds 4 to the *y*-coordinate (since $-2 + 4 = 2$). Therefore, the formula for the translation is: $(x, y) \rightarrow (x - 4, y + 4)$.

POE also works well here, because none of the other answer choices is even close to correct.

19. (4)

On a rather uncharacteristic question such as this one, you should definitely use POE and ask yourself, "Which of these things just doesn't belong here?" A rhombus is a quadrilateral with four equal sides, like this:

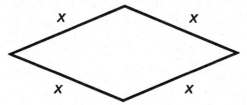

Run down the list of answer choices: (1) all rhombuses (also known as "rhombi") are quadrilaterals; (2) they all have a pair of equal adjacent sides (in fact, *every* pair of adjacent sides are of equal length); and (3) they're all parallelograms. However, a rhombus does not have to have any right angles, so to prove that a quadrilateral was a rectangle wouldn't help you prove it was a rhombus. By POE, answer choice (4) is the best choice.

20. (2)

Here's another question about rhombi. First, draw your picture:

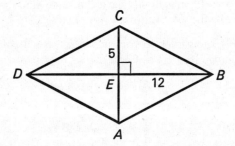

Focus your attention on △*EBC*. It's important to realize that the diagonals of a rhombus bisect each other and are perpendicular to each other. Therefore, side *BC* is the hypotenuse of right triangle *EBC*, whose legs measure 5 and 12. By now, you should recognize this as a 5:12:13 triangle (if not, use the Pythagorean Theorem), so *BC* = 13.

21. (3)

This problem requires a lot of work and tests several geometric principles. You have to test each answer choice, like so:

(1) Using the distance formula, calculate the distances AB and BC

$$AB = \sqrt{(x_2 - x_1)^2 + (y_2 - y_1)^2} \qquad BC = \sqrt{(x_2 - x_1)^2 + (y_2 - y_1)^2}$$
$$= \sqrt{(4-1)^2 + (5-2)^2} \qquad\qquad = \sqrt{(7-5)^2 + (6-4)^2}$$
$$= \sqrt{(3)^2 + (3)^2} \qquad\qquad\quad = \sqrt{(2)^2 + (2)^2}$$
$$= \sqrt{9+9} \qquad\qquad\qquad\quad = \sqrt{4+4}$$
$$= \sqrt{18} \qquad\qquad\qquad\qquad = \sqrt{8}$$

The distances are not equal, so cross off answer choice (1)

(2) Now find the slopes of AB and BC:

$$m(AB) = \frac{y_2 - y_1}{x_2 - x_1} \qquad\qquad m(BC) = \frac{y_2 - y_1}{x_2 - x_1}$$
$$= \frac{5-2}{4-1} \qquad\qquad\qquad = \frac{7-5}{6-4}$$
$$= \frac{3}{3} = 1 \qquad\qquad\qquad = \frac{2}{2} = 1$$

These two line segments have the same slope, so they're not perpendicular. Get rid of answer choice (2).

Use this knowledge to your advantage when you evaluate answer choice (3):

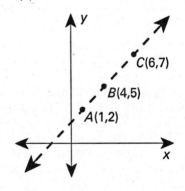

Since AB and BC have the same slope, they combine to form segment $\overline{AC}$, which has a slope of 1. The three points are collinear.

22. (4)

Before you do anything, find the measure of each of the angles:

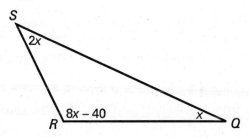

Use the Rule of 180:

$$m\angle Q + m\angle R + m\angle S = 180$$
$$x + (8x - 40) + 2x = 180$$
$$11x - 40 = 180$$
$$11x = 220$$
$$x = 20$$

From this information, you can figure out that $m\angle Q = 20$, $m\angle R = 120$, and $m\angle S = 40$. Right now is a good time to check that the three angles add up to 180°, which they do.) Since $m\angle R > 90$, $\angle R$ is an obtuse angle and $\triangle QRS$ is an obtuse triangle

23. (3)

Use one of De Morgan's Laws to solve this one:

$$\sim(a \wedge b) \rightarrow \sim a \vee \sim b$$

This basically means than when you negate a parenthetical statement with a "$\wedge$" or "$\vee$" in it, negate each symbol and flip the symbol upside down. Therefore:

$$\sim(q \wedge \sim s) \rightarrow \sim q \vee \sim(\sim s)$$

Since $\sim(\sim s)$ is the same thing as s (because of the rule of double negation), you can rewrite the statement as: $\sim q \vee s$.

24. (3)

After a reflection in the x-axis, each x-coordinate remains the same and each y-coordinate is negated. In other words, $r_{\text{x-axis}}(x, y) \rightarrow (x, -y)$. Therefore, when point $P(3, -4)$ is reflected in the x-axis, the coordinates of the image of point P' are $(3, 4)$.

25. (1)

When two lines are perpendicular, the slopes of the two lines are negative reciprocals. That is, their product equals -1. Since the line $y = \frac{1}{3}x - 2$ is in standard form, you know that its slope is $\frac{1}{3}$. A line perpendicular to the line $y = \frac{1}{3}x - 2$ will have a slope of -3. All of the answer choices are also standard form, and the only one with a slope of -3 is answer choice (1).

26. (1)

Given the lengths of two sides of a triangle, the length of the third side has to be smaller than the sum of the other two sides and larger than their difference.

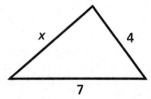

In this case, the length of the third side must be:

$$(7 - 4) < x < (7 + 4)$$
$$3 < x < 11$$

Only answer choice (1) is outside this range. Remember that the third side must be *less than* 11.

27. (2)

Use the formula for a circle, and remember that (h, k) is the center and r is the radius:

$$(x - h)^2 + (y - k)^2 = r^2$$
$$(x - 0)^2 + (y - 4)^2 = 3^2$$
$$x^2 + (y - 4)^2 = 9$$

Since the formula involves r^2 and not r, you should recognize that the formula will equal 9, not 3 or 0. Therefore, eliminate answer choices (1) and (3).

28. (1)

If you look at the diagonals of a parallelogram, a rectangle, a rhombus, and square, you'll find that the diagonals of each one bisect each other.

Think of it another way: all four quadrilaterals are parallelograms. Since the diagonals of a parallelogram bisect each other, it must be true for each of the quadrilaterals.

If something is certain to happen, then the probability that it will happen is 1.

29. (4)

When a point undergoes a dilation, each coordinate of that point is multiplied by a constant. Therefore, for the point (x, y) to be mapped onto the image point $(5x, 5y)$, it must undergo a dilation of 5.

30. (1)

Since $\triangle MNO$ and $\triangle QPO$ are right triangles, you might think of using Hypotenuse-Leg to prove that they are congruent. However, you don't know anything about the hypotenuse of either triangle. Therefore, using Hypotenuse-Leg is not useful.

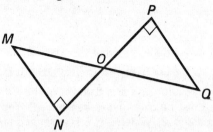

You know about two pairs of congruent sides and two pairs of congruent angles, so each of the other three answer choices is possible.

31. (4)

You can find the axis of symmetry of a parabola written in standard form ($y = ax^2 + bx + c$) by using the formula $x = -\dfrac{b}{2a}$. For this parabola, $a = 2$ and $b = -5$:

$$x = -\frac{(-5)}{2(2)} = \frac{5}{4}$$

Be careful with your minus signs, or you might get answer choice (3) instead.

32. (3)

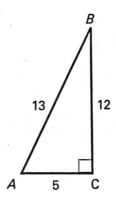

You're working with cosine (the CAH in SOHCAHTOA). Cosine equals adjacent over hypotenuse; the side adjacent to $\angle A$ is 5, and the hypotenuse is 13. Therefore, $\cos A = \dfrac{5}{13}$.

33. (3)

First, think to yourself about what the first statement means:

$\sim a \vee \sim b$ means "Either a is false OR b is false."

The second statement says b is true; from the Law of Disjunctive Inference, it must be true that a is false ($\sim a$). You can now eliminate answer choice (2), which says b is false, and answer choice (4), which says a is true.

Now use the Law of Contrapositive Inference (Flip-and-Negate) on the third statement:

$c \rightarrow a$ becomes $\sim a \rightarrow \sim c$. ("If a is false, then c is false.") You know a is false, so c must also be false.

34. (2)

This problem also uses the Law of Contrapositive Inference (Flip-and-Negate). First, symbolize the statements:

S represents "You are part of the solution."

P represents "You are part of the problem."

The statement in the problem now looks like this: ~S → P.

After you use the contrapositive, the statement looks like this: ~P → S. This is what is expressed in answer choice (2).

35. Construction

First, make sure the width of your compass is smaller than the length of $\overline{BC}$. Then put the pointy end on B and make an arc that intersects $\overline{BC}$ and $\overline{AB}$. Label those points M and P. Without changing the width of your compass, put the pointy end on W and make another arc like this:

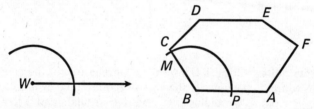

Now adjust the compass so that it's exactly as wide as $\overline{MP}$. Place the compass on X and make an arc that intersects with the first arc you drew from point W. Label the point of intersection Y:

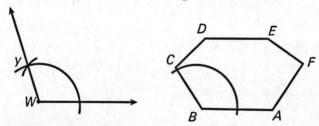

Connect ray WY.

Part II

36. *a*

Figure out the points you have to graph by plugging in all the integers between –1 and 5, inclusive. For example, if $x = -1$, then $y = (-1)^2 - 4(-1) + 3$, or 8. Your first coordinate is $(-1, 8)$. The rest of your T-chart is below, along with the graph of the parabola:

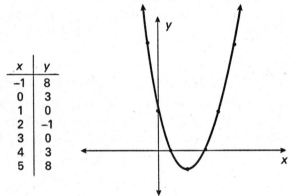

x	y
–1	8
0	3
1	0
2	–1
3	0
4	3
5	8

b

A translation "slides" a figure to a different area of the coordinate axes without changing its shape. Use the formula given to find out the new points by adding 3 to each x-coordinate and 2 to each y-coordinate. The new T-chart and sketch are below:

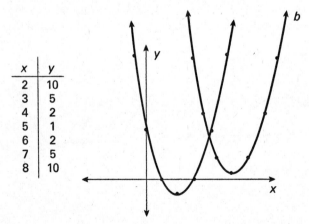

x	y
2	10
3	5
4	2
5	1
6	2
7	5
8	10

c

A reflection in the x-axis maps the point (x, y) onto the image point $(x, -y)$. You have to make one more list of points, and your graph should look like this (remember to put all three parabolas on the same set of coordinate axes and refer to each one neatly and specifically):

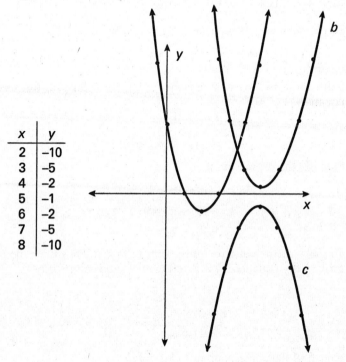

x	y
2	−10
3	−5
4	−2
5	−1
6	−2
7	−5
8	−10

37. a $\dfrac{4}{x+3}$

Factor the algebraic terms in the fractions like this:

$$4x - 24 = 4(x - 6)$$
$$x^2 - 36 = (x + 6)(x - 6)$$
$$x^2 + 4x - 12 = (x + 6)(x - 2)$$
$$x^2 + x - 6 = (x + 3)(x - 2)$$

Now rewrite the problem like this, and cancel out all the terms that appear both on the top and on the bottom:

$$\frac{4(x-6)}{(x+6)(x-6)} \cdot \frac{(x+6)(x-2)}{(x+3)(x-2)}$$

$$\frac{4(x-6)}{(x+6)(x-6)} \cdot \frac{(x+6)(x-2)}{(x+3)(x-2)} = \frac{4}{x+3}$$

To check your work, you can plug in a value for x.

b (–9, –4) and (4, 9)

Solve these two systems by substituting $x + 5$ for y in the second equation:

$$y = x + 5$$
$$x^2 + y^2 = 97$$
$$x^2 + (x+5)^2 = 97$$

Simplify the expression:

$$x^2 + (x^2 + 10x + 25) = 97$$
$$2x^2 + 10x + 25 = 97$$
$$2x^2 + 10x - 72 = 0$$

To make this equation a little easier to factor, you can divide each term in the equation by 2. Then factor it and solve for x:

$$x^2 + 5x - 36 = 0$$
$$(x+9)(x-4) = 0$$
$$x = \{-9, 4\}$$

Now find the corresponding values of y:

If $x = -9$, then $y = -9 + 5$; $y = -4$.   Solution One: (–9, –4).

If $x = 4$, then $y = 4 + 5$; $y = 9$.   Solution Two: (4, 9).

38. 5.4

The first thing to realize is that $\triangle BDE$ and $\triangle BAC$ are similar. Since $\overline{DE}$ is parallel to $\overline{AC}$, it must be true that $\angle BDE \cong \angle A$ and $\angle BED \cong \angle C$ (both pairs are corresponding angles, so they're congruent to each other). The triangles are similar because of the Angle-Angle Similarity Theorem.

Now you can set up proportions between corresponding sides of the two triangles. First, though, you have to assign values to them.

Let $AD = x$. Since DB is 2 less than AD, let $DB = x - 2$. Also, $AC = x + 5$ (because AC is 5 more than AD). Label the sides like this:

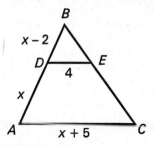

Your proportion should look like this:

$$\frac{DB}{AB} = \frac{DE}{AC}$$

Fill in the algebraic terms you've just figured out, and remember that $AB = AD + DB$:

$$\frac{x - 2}{x + (x - 2)} = \frac{4}{x + 5}$$

Cross-multiply and solve:

$$(x - 2)(x + 5) = 4(2x - 2)$$
$$x^2 + 3x - 10 = 8x - 8$$
$$x^2 - 5x - 2 = 0$$

You can't factor this, so you have to use the Quadratic Formula:

$$x = \frac{-b \pm \sqrt{b^2 - 4ac}}{2a}$$

In the equation $x^2 - 5x - 2 = 0$, $a = 1$, $b = -5$, and $c = -2$:

$$x = \frac{-(-5) \pm \sqrt{(-5)^2 - 4(1)(-2)}}{2(1)} = \frac{5 \pm \sqrt{25 + 8}}{2} = \frac{5 \pm \sqrt{33}}{2}$$

At this point, use your calculator to find out that $\sqrt{33} \approx 5.74$. When you replace $\sqrt{33}$ with 5.74, you have two possible answers:

$$x = \frac{5+5.74}{2} = \frac{10.74}{2} = 5.37 \text{ or } x = \frac{5-5.74}{2} = \frac{-0.74}{2} = -0.37$$

Since your second option is negative, it's not a possible answer (you're looking for the length of something, so your answer must be positive). Therefore, your only answer is 5.37, which rounds to 5.4. (Remember to do as you were instructed and round to the nearest *tenth*, or you'll lose points.)

39. ***a*** **(1) 121**

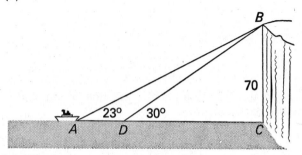

Luckily, the diagram they gave you is wonderfully explanatory. Even though this problem looks like something you'd have to know to get into the Naval Academy, it's only a simple trigonometry problem (with two right angles).

DC is the length of the leg adjacent to $\angle BDC$ in $\triangle BDC$. You know the opposite side and the adjacent side of the angle, so use tangent (the TOA in SOHCAHTOA):

$$\tan \angle BDC = \frac{70}{DC}$$

$$\tan 30° = \frac{70}{DC}$$

$$0.5774 = \frac{70}{DC}$$

$$0.5774(DC) = 70$$

$$DC = \frac{70}{0.5774}$$

$$DC \approx 121.23$$

Rounded off to the nearest *meter*, the answer is 121 meters.

(2) 165

Follow a similar process to find AC, which is the length of the adjacent leg of $\triangle BAC$:

$$\tan 23° = \frac{70}{AC}$$

$$0.4245 = \frac{70}{AC}$$

$$0.4245(AC) = 70$$

$$AC = \frac{70}{0.4245}$$

$$AC \approx 164.89$$

Rounded off to the nearest *meter*, the answer is 165 meters.

(3) 179

This doesn't require any trig at all. AB is the length of the hypotenuse of $\triangle BAC$, and you know the length of both legs:

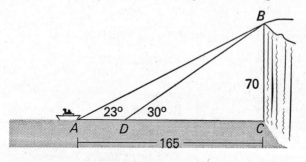

Use the Pythagorean Theorem:

$$(AB)^2 = (AC)^2 + (BC)^2$$

$$(AB)^2 = (165)^2 + (70)^2$$

$$(AB)^2 = 27,225 + 4,900$$

$$(AB)^2 = 32,125$$

$$AB \approx 179.23$$

This rounds off to 179 meters.

b **44**

If you've done your math correctly so far, this part is a snap. The difference between the two positions of the ship is represented by AD. Since $AD + DC = AC$, you can plug in the distances you already know:

$$AD + 121 = 165$$
$$AD = 44 \text{ meters}$$

40.

In this proof, the test has already mapped out the symbols.

Step One: Turn all the givens into symbolic terms:

"If Ronnie does not waste time in class, then she does well in Course II." $\sim T \rightarrow B$

"If Ronnie is absent from class, then her grades will go down."

$A \rightarrow S$

"Either Ronnie does not waste time in class or Ronnie is absent from class." $\sim T \vee A$

"Ronnie's grades do not go down." $\sim S$

Step Two: Decide what you want to prove:

"Ronnie does well in Course II." B

Step Three: Write the proof.

Statements	Reasons
1. $A \rightarrow S$; $\sim S$	1. Given
2. $\sim A$	2. Law of *Modus Tollens*
3. $\sim T \vee A$	3. Given
4. $\sim T$	4. Law of Disjunctive Inference (2, 3)
5 $\sim T \rightarrow B$	5. Given
6. B	6. Law of Detachment (4, 5)

Part III

41.

The plan: This looks like those problems involving an altitude drawn within a right triangle that creates three similar triangles. Show that $\triangle AEC$ and $\triangle BDA$ are similar, which will prove that the corresponding sides are proportional.

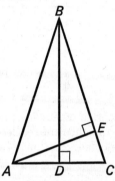

Statements	Reasons
1. $\overline{BA} \cong \overline{BC}$	1. Given
2. $\angle BAD \cong \angle ACE$	2. If two sides of a triangle are congruent, then the angles opposite those sides are congruent.
3. $\overline{AE} \perp \overline{BC}$; $\overline{BD} \perp \overline{AC}$	3. Given
4. $\angle AEC$ and $\angle BDA$	4. Definition of a right angle are right angles
5. $\angle AEC \cong \angle BDA$	5. All right angles are congruent.
6. $\triangle AEC \cong \triangle BDA$	6. AA rule for Similarity
7. $\dfrac{AC}{BA} = \dfrac{AE}{BD}$	7. The lengths of corresponding sides of similar triangles are in proportion.

42.

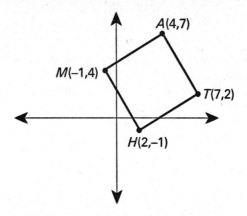

A square is a parallelogram with at least two adjacent sides that are equal in length and perpendicular to each other. To prove that *MATH* is a parallelogram, calculate the slope of each side using the slope formula, which is $m = \dfrac{y_2 - y_1}{x_2 - x_1}$:

$$m(MA) = \frac{7 - 4}{4 - (-1)} = \frac{3}{5} \qquad m(AT) = \frac{2 - 7}{7 - 4} = \frac{-5}{3} - \frac{5}{3}$$

$$m(TH) = \frac{-1 - 2}{2 - 7} = \frac{-3}{-5} = \frac{3}{5} \qquad m(MH) = \frac{-1 - 4}{2 - (-1)} = \frac{-5}{3} = -\frac{5}{3}$$

Since opposite sides $\overline{MA}$ and $\overline{TH}$ have the same slope, the two segments are parallel. The same is true of sides $\overline{AT}$ and $\overline{MH}$; since the opposite sides of *MATH* are parallel, *MATH* is a parallelogram.

Now consider adjacent sides $\overline{MA}$ and $\overline{AT}$. If you can show that they are perpendicular to each other and have the same length, you can prove that *MATH* is a square. The first part is easy; the slope of $\overline{MA}$ is $\dfrac{3}{5}$, and the slope of $\overline{AT}$ is $-\dfrac{5}{3}$. Since these slopes are negative reciprocals $\left(\dfrac{3}{5} \times -\dfrac{5}{3} = -1\right)$, $\overline{MA}$ and $\overline{AT}$ are perpendicular

Using the distance formula, calculate the distances MA and AT.

$$MA = \sqrt{(x_2 - x_1)^2 + (y_2 - y_1)^2}$$
$$= \sqrt{[4 - (-1)]^2 + (7 - 4)^2}$$
$$= \sqrt{5^2 + 3^2}$$
$$= \sqrt{25 + 9}$$
$$= \sqrt{34}$$

$$AT = \sqrt{(x_2 - x_1)^2 + (y_2 - y_1)^2}$$
$$= \sqrt{(7 - 4)^2 + (2 - 7)^2}$$
$$= \sqrt{3^2 + (-5)^2}$$
$$= \sqrt{9 + 25}$$
$$= \sqrt{34}$$

Quadrilateral $MATH$ is a parallelogram with at least one pair of sides that are congruent and perpendicular to each other. Therefore, $MATH$ is a square.

EXAMINATION:
JUNE 1994

Part I

Answer 30 questions from this part. Each correct answer answer will receive 2 credits. No partial credit will be allowed. Write your answers in the spaces provided on the separate answer sheet. Where applicable, answers may be left in terms of π or in racial form. [60]

1 Segment $\overline{AB}$ is parallel to segment $\overline{CD}$. If the slope of $\overline{AB} = -\dfrac{3}{7}$ and the slope of $\overline{CD} = -\dfrac{x}{14}$, find the value of x.

2 Lines $\overleftrightarrow{AB}$ and $\overleftrightarrow{CD}$ intersect at point F. What is the total number of points 4 centimeters from point F and also equidistant from $\overleftrightarrow{AB}$ and $\overleftrightarrow{CD}$?

3 In the following system, determine the value of $(a \odot b) \odot c$.

$\odot$	a	d	b	c
a	b	a	c	d
d	a	d	b	c
b	c	b	d	a
c	d	c	a	b

4 If a translation maps $(x,y) \rightarrow (x + 2, y + 3)$, what are the coordinates of B', the image of point $B(-3,5)$ after this translation?

5 In the accompanying diagram, $l \parallel m$, t and s are intersecting transversals, $\angle 1 = 130$, and m $\angle 2 = 60$. Find m $\angle 3$.

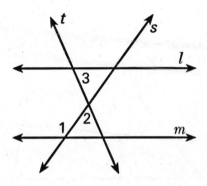

6 In $\triangle ABC$, m $\angle A = 65$ and m $\angle C = 60$. Which is the *shortest* side of the triangle?

7 If $\tan A = \dfrac{3}{4}$, find m $\angle A$ to the *nearest degree*.

8 In the accompanying diagram, the altitude to the hypotenuse of the right triangle divides the hypotenuse into two segments of lengths 3 and 12. What is the length of the altitude?

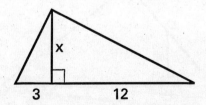

9 What are the coordinates of P', the image of $P(1,2)$ after a reflection in the origin?

10 The coordinates of A and B are $(2a,4b)$ and $(8a,6b)$, respectively. Express, in terms of a and b, the coordinates of the midpoint of $\overline{AB}$.

11 In isosceles triangle ABC, $\overline{AB} \cong \overline{CB}$. Find m$\angle B$, if m$\angle A = 5x - 4$ and m$\angle C = 2x + 20$.

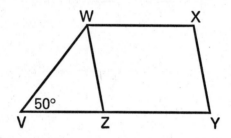

12 In the accompanying diagram, WYXZ is a parallelogram, line $\overline{YZ}$ is extended to point V, $\overline{WZ} \cong \overline{VZ}$, and m$\angle V = 50$. Find m$\angle ZWX$.

13 In $\triangle ABC$, $\overline{AB} \perp \overline{BC}$, and $\overline{DE} \perp \overline{CA}$. If $DE = 8$, $CD = 10$, and $CA = 30$, find AB.

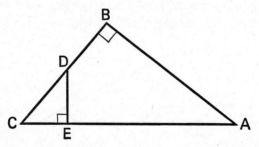

14 Write an equation of the line that passes through points (2,3) and (4,5).

15 What is the positive root of the equation $c^2 - 6c = 27$?

16 If the length of one side of a rectangle is 8 and the length of a diagonal is 10, find the area of the rectangle.

Directions (17-34): For *each* question chosen, write the *numeral* preceding the word or expression that best completes the statement or answers the question.

17 Two consecutive angles of a parallelogram measure $2x + 10$ and $x - 10$. What is the value of x?

(1) 30 (3) 120
(2) 60 (4) –20

18 What is the length of the line segment joining points $J(1,5)$ and $K(3,9)$?

(1) $2\sqrt{5}$ (3) $13\sqrt{2}$
(2) $\sqrt{13}$ (4) $2\sqrt{13}$

19 Which statement is logically equivalent to $\sim(a \wedge \sim b)$?

(1) $\sim a \wedge b$ (3) $\sim a \vee b$
(2) $\sim a \wedge \sim b$ (4) $\sim a \vee b$

20 Which statement is equivalent to the inequality $9 - 4x \leq 3x - 5$?

(1) $x > -2$ (3) $x \leq -2$
(2) $x < 2$ (4) $x \geq 2$

21 Which polygon must have congruent diagonals?

(1) parallelogram   (3) trapezoid
(2) rectangle   (4) rhombus

22 What is the y-intercept of the graph of the equation $y = 2x^2 - 5x + 7$?

(1) -5 (3) 7
(2) 2 (4) -7

23 If the statements $m \rightarrow n$ and $\sim m \rightarrow s$ are true, then which statement is a logical conclusion?

(1) $n \rightarrow s$ (3) s
(2) $s \rightarrow n$ (4) $\sim n \rightarrow s$

24 Which equation describes the locus of points 5 units from point $(3,-4)$?

(1) $(x + 3)^2 + (y - 4)^2 = 5$
(2) $(x - 3)^2 + (y + 4)^2 = 5$
(3) $(x - 3)^2 + (y + 4)^2 = 25$
(4) $(x + 3)^2 + (y - 4)^2 = 25$

25 In the solution of this problem, which property of real numbers justifies stament 5?

Statements	Reasons
1. $3x = 6$	1. Given
2. $\frac{1}{3}(3x) = \frac{1}{3}(6)$	2. Multiplication axiom
3. $\left(\frac{1}{3} \cdot 3\right)x = 2$	3. Associative property
4. $1 \times x = 2$	4. Multiplicative inverse
5. $x = 2$	5. _____

(1) Closure   (3) Communicative
(2) Identity   (4) Inverse

26 How many 9-letter arrangements can be formed from the letters in the word "SASSAFRAS"?

(1) $\dfrac{4!}{3!}$ (3) $\dfrac{9!}{7!}$

(2) $\dfrac{9!}{4!\,3!}$ (4) $9!$

27 If the length of each leg of an isosceles triangle is 17 and the base is 16, the length of the altitude to the base is

(1) 8   (3) 15
(2) $8\dfrac{1}{2}$ (4) $\sqrt{32}$

28 Which equation represents the line that passes through point (0,6) and is perpendicular to the line whose equation is $y = 3x - 2$?

(1) $y = -\dfrac{1}{3}x + 6$ (3) $y = -3x + 6$

(2) $y = \dfrac{1}{3}x + 6$ (4) $y = 3x + 6$

29 Expressed as a fraction in lowest terms, $\dfrac{1}{x^2 - 4} \div \dfrac{x}{x - 2}$, $x \neq 2, 0, -2$, is equivalent to

(1) $\dfrac{1}{x(x + 2)}$ (3) $\dfrac{1}{x(x - 2)}$

(2) $\dfrac{-2}{x^2 - 4}$ (4) $\dfrac{1}{2}$

30 The lengths of two sides of a triangle are 7 and 10. The length of the third side may be
(1) 17   (3) 3
(2) 20   (4) 8

31 Which expression is not equivalent to $_7C_5$?

(1) $_7P_5$ (3) $\dfrac{7 \cdot 6 \cdot 5 \cdot 4 \cdot 3}{5 \cdot 4 \cdot 3 \cdot 2 \cdot 1}$

(2) 21   (4) $_7C_2$

32 What are the roots of the equation $2x - 6x + 3 = 0$?

(1) $\dfrac{-3 \pm \sqrt{3}}{2}$ (3) $\dfrac{3 \pm \sqrt{3}}{2}$

(2) $\dfrac{-3 \pm \sqrt{15}}{2}$ (4) $\dfrac{3 \pm \sqrt{15}}{2}$

33 The sum of $\dfrac{x+4}{x}$ and $\dfrac{x-4}{4}$ is

(1) $\dfrac{1}{2}$ (3) $\dfrac{x^2+16}{4x}$

(2) $4+x$ (4) $\dfrac{2x}{x+4}$

34 In the accompanying diagram, $\triangle ABC$ is a scalene triangle.

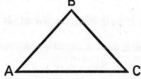

If the median is drawn from vertex B, what is the probability that its length will be greater than the length of the altitude?

(1) 1 (3) $\dfrac{1}{2}$

(2) 0 (4) $\dfrac{3}{4}$

Directions (35): Show all construction lines.

35 On the answer sheet, construct the ray that bisects $\angle B$.

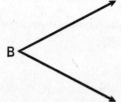

Part II

Answer *three* questions from this part. Clearly indicate the necessary steps, including appropriate formula substitutions, diagrams, graphs, charts, etc. Calculations that may be obtained by mental arithmetic or the calculator do not need to be shown. [40]

36 Answer both *a* and *b* for all values of *x* for which these expressions are defined.

   a Express the product in simplest form:

$$\frac{x^2 - 9}{x^2 - x - 20} \bullet \frac{4x^2 - 20x}{4x^2 - 12x} \quad \text{[6]}$$

   b Solve for *x*: $\dfrac{x - 3}{2} = \dfrac{6}{x + 8}$ [4]

37 *a* On graph paper, draw the graph of the equation $y = x^2 - 4x + 4$, including all values of *x* from $x = -1$ to $x = 5$. Label the graph *a*. [4]

   b On the same set of axes, draw the image of the graph drawn in part *a* after a translation that maps $(x,y) \rightarrow (x - 2, y + 3)$. Label the image *b*. [2]

   c On the same set of axes, draw the image of the graph drawn in part *b* after a reflection in the *x*-axis. Label the image *c*. [2]

   d Which equation could represent the graph drawn in part *c*? [2]

   (1) $y = -x^2 + 4x - 4$
   (2) $y = x^2 - 3$
   (3) $y = -x^2 - 3$
   (4) $y = -x^2 + 3$

38 Alan has three detective books, two books about cars, and five comic books. He plans to lend three books to his friend David.

a How many different selections of three books can Alan lend his friend? [2]

b Find the probability that a three-book selection will contain
  (1) one book of each type [3]
  (2) comic books, only [3]
  (3) books about cars, only [2]

39 Trapezoid *ABCD* has coordinates *A*(–6,0), *B*(17,0), *C*(2,8), *D*(0,8). Find the

a area of trapezoid *ABCD* [3]

b perimeter of *ABCD* [4]

c measure of ∠*B* to the *nearest degree* [3]

40 *On your answer paper*, write the numerals 1 through 8, and next to each numeral, give a reason for each statement in the proof. For statement 1, write "Given."

Given: $\triangle ABC$, $\overline{AC} \cong \overline{BC}$, $\overline{AD}$ and $\overline{BE}$ intersect at G, and $\angle 1 \cong \angle 2$.

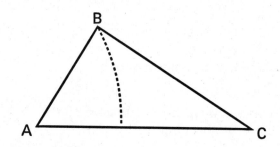

Prove: $\overline{EG} \cong \overline{DG}$

Statements	Reasons	
(1) $\triangle ABC$, $\overline{AC} \cong \overline{BC}$, $\angle 1 \cong \angle 2$	(1) Given	
(2) $CAB\ CBA$	(2)	[2]
(3) $\overline{AB} \cong \overline{BA}$	(3)	[1]
(4) $\triangle EAB \cong \triangle DBA$	(4)	[2]
(5) $\angle AEB \cong \angle BDA$, $\overline{AE} \cong \overline{BD}$	(5)	[1]
(6) $\angle EGA \cong \angle DGB$	(6)	[1]
(7) $\triangle EGA \cong \triangle DGB$	(7)	[2]
(8) $\overline{EG} \cong \overline{DG}$	(8)	[1]

Part III

Answer one question from this part. Clearly indicate the necessary steps, including appropriate formula substitutions, diagrams, graphs, charts, etc. Calculations that may be obtained by mental arithmetic or the calculator do not need to be shown. [10]

41 Given: If pro basketball players compete in the Olympics, then college players do not play.

If college players do not play, then the team is not an amateur team.

If the team is not an amateur team and the team does not win the gold medal, then the people are not happy.

Pro basketball players compete in the Olympics.

The people are happy.

Let *P* represent: "Pro basketball players compete in the Olympics."

Let *C* represent: "College players play."

Let *A* represent: "The team is an amateur team."

Let *G* represent: "The team wins the gold medal."

Let *H* represent: "The people are happy."

42 The coordinates of the vertices of *TAG* are $T(1,3)$, $A(8,2)$, and $G(5,6)$. Prove that $\triangle TAG$ is an isosceles right triangle. [10]

ANSWER KEY

Part I

1. 6

2. 4

3. b

4. (−1,8)

5. 70

6. $\overline{AC}$

7. 37

8. 6

9. (−1,−2)

10. (5a,5b)

11. 108

12. 80

13. 24

14. $y=x + 1$

15. 9

16. 48

17. (2)

18. (1)

19. (4)

20. (4)

21. (2)

22. (3)

23. (4)

24. (3)

25. (2)

26. (2)

27. (3)

28. (1)

29. (1)

30. (4)

31. (1)

32. (3)

33. (3)

34. (1)

35. construction

EXPLANATIONS:
JUNE 1994

Part I

1. 6

If two lines (or line segments) are parallel, they have the same slope.
Set the slope of $\overline{AB}$ equal to the slope of $\overline{CD}$. Place the minus sign
in the numerator of each fraction and cross-multiply:

$$\frac{-3}{7} = \frac{-x}{14}$$
$$-7x = -42$$
$$x = 6$$

2. 4

The set of points that are 4 centimeters from point F is a circle with
a radius of 4 cm centered at point F:

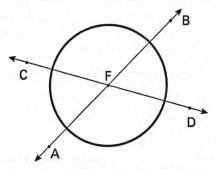

The set of points equidistant from lines AB and CD is a pair of lines
l and m (see below), each of which is an angle bisector. Line l bisects $\angle AFC$ and $\angle DFB$, and line m bisects $\angle AFD$ and $\angle CFB$.

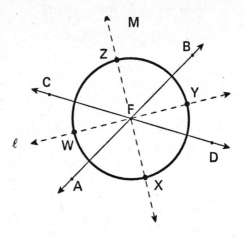

The two sets of points intersect in four points: W, X, Y, and Z.

3. b

This is a two-part system question, which means that you have to follow system directions twice instead of only once. You have to find the value of $(a \odot b)$ first; to do so, find a on the left side of the box and follow along the horizontal row until you hit the column with the b on top. That value is c, so $(a \odot b) = c$.

Now find $(c \odot c)$ the same way. The c column intersects the c row at the value b.

4. (–1, 8)

If a translation maps the point (x, y) onto the point $(x + 2, y + 3)$, that means you add 2 to each x-coordinate and 3 to each y-coordinate. The image of $B(-3, 5)$ is therefore $B'(-3 + 2, 5 + 3)$, or $B'(-1, 8)$.

5. 70

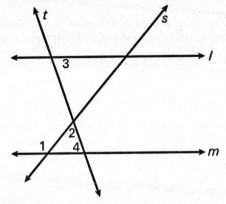

Since the measure of an exterior angle equals the sum of the measures of the two non-adjacent interior angles of the triangle, it must be true that m ∠1 = m ∠2 + m ∠4. Using the data you know, solve for m ∠4:

$$130 = 60 + m\angle 4$$
$$70 = m\angle 4$$

Since lines l and m are parallel, ∠3 and ∠4 are alternate interior angles, which have the same measure. Therefore, m ∠3 = m ∠4 = 70.

6. $\overline{AC}$

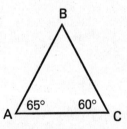

To figure out the measure of ∠B, use the Rule of 180:

$$m\angle A + m\angle B + m\angle C = 180$$
$$65 + m\angle B + 60 = 180$$
$$m\angle B = 55$$

The shortest side of the triangle is opposite the smallest angle. Since ∠B is the smallest angle, the side opposite it, $\overline{AC}$, is the smallest side.

7. 37

This is a job for your scientific calculator. Since the question asks for the answer in degrees, make sure your calculator is in degree mode. If tan $A = \dfrac{3}{4}$, then the measure of angle A is the inverse tangent (or tan^{-1}) of $\dfrac{3}{4}$. Convert $\dfrac{3}{4}$ to a decimal (0.75), then enter it into your calculator. Now find the "tan^{-1}" button on your calculator (it's usually above the "tan" button—you may have to press the "second function" button) and press it. You should get 36.87, which rounds up to 37°.

8. 6

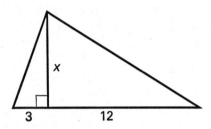

This figure represents three similar right triangles, and of all their corresponding sides are proportional to each other. For this problem, consider the small one and the medium one, and set up a proportion. In the small one, the short leg is 3 and the long leg is x. In the medium triangle, the short leg is x and the long leg is 12. Set up the proportion:

$$\frac{3}{x} = \frac{x}{12}$$

When you cross-multiply, you'll get:

$$x(x) = 3 \times 12$$
$$x^2 = 36$$
$$x = \{6, -6\}$$

Since you're looking for a distance, which can't be a negative value, eliminate –6. The value of x is 6.

9. **(–1,–2)**

The formula for the reflection of a point (x, y) in the origin is $P(–x, –y)$. Therefore, the image of the point $(1, 2)$ reflected in the origin is $(–1, –2)$.

10. **($5a$, $5b$)**

The formula for the midpoint of a line segment is:

$$(\overline{x}, \overline{y}) = \left(\frac{x_1 + x_2}{2}, \frac{y_1 + y_2}{2} \right)$$

Even though the points you've been given look a little unorthodox, you can plug them into the formula just as you would plug in regular numbers:

$$(\overline{x}, \overline{y}) = \left(\frac{2a + 8a}{2}, \frac{4b + 6b}{2} \right) = \left(\frac{10a}{2}, \frac{10b}{2} \right) = (5a, 5b)$$

11. **108**

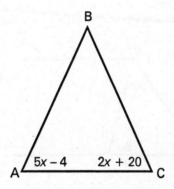

Since $\overline{AB} \cong \overline{CB}$, the angles opposite them must also be equal. Set the measures of $\angle A$ and $\angle C$ equal to each other:

$$5x - 4 = 2x + 20$$
$$3x = 24$$
$$x = 8$$

Now, plug 8 in for x to find the measure of each angle (it pays to plug into both of them to make sure you did your math right):

$$m \angle A = 5(8) - 4 = 36$$
$$m \angle C = 2(8) + 20 = 36$$

Now use the Rule of 180 to figure out the measure of $\angle B$:

$$m\angle A + m\angle B + m\angle C = 180$$
$$36 + m\angle B + 36 = 180$$
$$m\angle B = 108$$

12. 80

The key to solving this problem is $\triangle VWZ$. Since $\overline{WZ} \cong \overline{VZ}$, $\triangle VWZ$ is isosceles and the two angles opposite the congruent sides ($\angle V$ and $\angle VWZ$) have equal measures. Thus, $m\angle VWZ = 50$.

From there, use the Rule of 180 to find the measure of $\angle VZW$:

$$m\angle V + m\angle VWZ + m\angle VZW = 180$$
$$50 + 50 + m\angle VZW = 180$$
$$m\angle VZW = 80$$

Next, look at sides $\overline{WX}$ and $\overline{ZY}$ of the parallelogram. Since opposite sides of a parallelogram are parallel, $\angle VZW$ and $\angle ZWX$ are alternate interior angles (which always have the same measure):

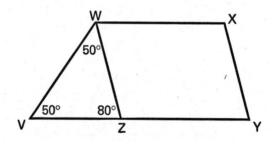

Thus, $m\angle VZW = m\angle ZWX = 80$.

13. 24

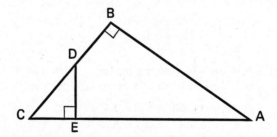

To answer this question, you have to do some geometric figuring. Consider $\triangle ABC$ and $\triangle DEC$. Since $\angle B \cong \angle DEC$ (because all right angles are congruent to one another) and $\angle C \cong \angle C$ (because all angles are equal to themselves—otherwise known as the reflexive property), the two right triangles are similar (because of the Angle-Angle Rule of Similarity).

Now, you know that the lengths of corresponding sides of the two triangles are proportional. CA and CD are hypotenuses, and AB and DE are the larger legs. Set up the following proportion:

$$\frac{CA}{CD} = \frac{AB}{DE}$$
$$\frac{30}{10} = \frac{AB}{8}$$

Cross-multiply, and you're done:

$$10 \times (AB) = 30 \times 8$$
$$10(AB) = 240$$
$$AB = 24$$

14. $y = x + 1$

Your final answer will be in $y = mx + b$ format, so it pays to find the slope of the line using the slope formula:

$$m = \frac{y_2 - y_1}{x_2 - x_1} = \frac{5 - 3}{4 - 2} = \frac{2}{2} = 1$$

The slope of the line is 1, so the equation so far is $y = (1)x + b$, or just $y = x + b$. Now you have to find the y-intercept (the b) by plugging the coordinates of one of the points into the equation. Choose $(2, 3)$:

$$3 = 2 + b$$
$$1 = b$$

The equation of the line is $y = x + 1$.

15. 9

Before you factor this, you have to alter the equation slightly until it equals zero:

$$c^2 - 6c = 27$$
$$c^2 - 6c - 27 = 0$$

Now use your best trial-and-error technique to factor it:

$$(c - 9)(c + 3) = 0$$
$$c = \{9, -3\}$$

They want the positive root, so toss out –3. Check your work to make sure it's right.

16. 48

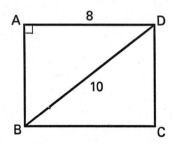

The key is to find the height of the rectangle (*AB*). Since a rectangle has four right angles, $\triangle ABD$ is a right triangle with one leg of length 8 and a hypotenuse of length 10. You might be able to recognize this as a 3:4:5 triangle—the dimensions are 6:8:10. (If you can't see this, use the Pythagorean Theorem.)

Since *AB* = 6, you now have the rectangle's dimensions. The area of a rectangle equals the length times the width, so the area of rectangle *ABCD* is 6 × 8, or 48.

Multiple Choice

17. (2)

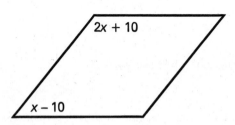

The sum of any two consecutive angles in a parallelogram is 180°.

Therefore:

$$(2x + 10) + (x - 10) = 180.$$
$$3x = 180$$
$$x = 60$$

Check your work to make sure it's correct.

18. (1)

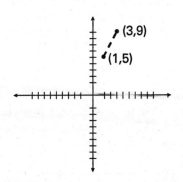

Use the distance formula:

$$d = \sqrt{(x_2 - x_1)^2 + (y_2 - y_1)^2} = \sqrt{(3 - 1)^2 + (9 - 5)^2}$$
$$= \sqrt{2^2 + 4^2} = \sqrt{4 + 16} = \sqrt{20}$$

This answer isn't among the four given, but don't panic. You have to simplify your answer by factoring out a perfect square:

$$\sqrt{20} = \sqrt{4 \times 5} = \sqrt{4} \times \sqrt{5} = 2\sqrt{5}$$

19. (4)

Use one of De Morgan's Laws to solve this one:

$$\sim(a \wedge b) \rightarrow \sim a \ \vee \sim b$$

This basically means that when you negate a parenthetical statement with a "∧" or "∨" in it, negate each term and flip the symbol upside down. Therefore:

$$\sim(a \ \wedge \sim b) \rightarrow \sim a \ \vee \sim(\sim b)$$

Since $\sim(\sim b)$ is the same thing as b (because of the rule of double negation), you can rewrite the statement as: $\sim a \vee b$.

20. (4)

Solve this inequality just as you would solve an equality. First, subtract $3x$ from both sides:

$$9 - 4x \leq 3x - 5$$
$$-7x \leq -14$$

Then divide both sides by -7. (Remember the important rule with inequalities: When you multiply or divide both sides by a negative number, flip the sign in the other direction!):

$$-7x \leq -14$$
$$x \geq 2$$

21. (2)

It's possible for each of the four quadrilaterals in the answer choices to have congruent diagonals. However, the only one that always has congruent diagonals is the rectangle:

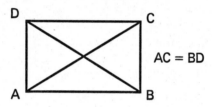

$$AC = BD$$

22. (3)

The y-intercept of a graph is the point at which it intersects the y-axis. At this point, the value of x is zero. The easiest way to find the y-intercept is to set $x = 0$ and solve for y:

$$y = 2(0)^2 - 5(0) + 7$$
$$y = 7$$

Note: When the formula of a parabola is in standard form $ax^2 + bx + c = 0$ (as this equation is), the y-intercept is the value of c. If you know this, you can look at the formula $y = 2x^2 - 5x + 7$ and realize right away that the y-intercept is 7.

23. (4)

You can't string these two logical statements together until they have a common term. The easiest way to combine them is to use the Law of Contrapositive Inference (the old Flip-and-Negate) on the first statement: $m \to n$ becomes $\sim n \to \sim m$.

Now you can link the two statements together using the Law of Syllogism (or Chain Rule). Given that $\sim n \to \sim m$ and $\sim m \to s$, it must be true that $\sim n \to s$.

24. (3)

The locus of points 5 units away from the point $(3, -4)$ is a circle with radius 5 centered at the point $(3, -4)$.

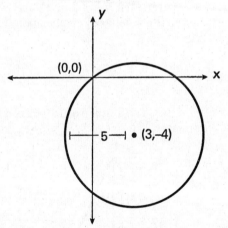

Use the formula for a circle, and remember that (h, k) is the center and r is the radius:

$$(x - h)^2 + (y - k)^2 = r^2$$
$$(x - 3)^2 + [y - (-4)]^2 = 5^2$$
$$(x - 3)^2 + (y + 4)^2 = 25$$

Since the formula involves r^2 and not r, you should recognize that the formula will equal 25, not 5. Therefore, eliminate answer choices (1) and (2).

25. (2)

1 is the multiplicative identity, which is a fancy way of saying that "anything times 1 equals itself." Therefore, the fact that $1 \times x = x$ is true is because of the multiplicative identity.

26. (2)

If SASSAFRAS didn't have any duplicate letters, answer choice (4) would be correct. If a word has n letters, and one letter is repeated p times and another is repeated q times (remember that p and q are greater than 1), then the number of possible arrangements of that word is:

$$\frac{n!}{p!\,q!}$$

SASSAFRAS has 9 letters, but it has four S's and three A's. Therefore, you can express the number of combinations as:

$$\frac{9!}{4!\,3!}$$

27. (3)

Draw your diagram first. The altitude of an isosceles triangle bisects the base of the triangle, like so:

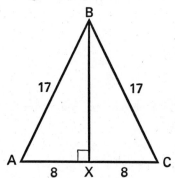

Now you can look at the altitude BX as the leg of a right triangle with one leg (AX) of length 8 and a hypotenuse (AB) of length 17, and you can use the Pythagorean Theorem:

$$(AX)^2 + (BX)^2 = (AB)^2$$
$$8^2 + (BX)^2 = 17^2$$
$$64 + (BX)^2 = 289$$
$$(BX)^2 = 225$$
$$BX = 15$$

Note: In the future, learn to recognize 8:15:17 as a Pythagorean Triplet. It's not very common, but it shows up every once in a while.

28. (1)

This problem appears to require a lot of math, but you really don't have to calculate much of anything. The line $y = 3x - 2$ is in $y = mx + b$ format, so the slope of the line must be 3. A line perpendicular to this line will have a slope that is the negative reciprocal of 3, or $-\dfrac{1}{3}$. The only line with a slope of $-\dfrac{1}{3}$ among the answer choices is answer choice (1).

Note: You don't even need to worry about the point (0,6)!

29. (1)

This looks similar to what you might find in Part Two of this exam (although it's not nearly as difficult). The process is the same. First, turn the problem into a multiplication exercise by flipping the second term:

$$\frac{1}{x^2 - 4} \div \frac{x}{x - 2}$$
$$= \frac{1}{x^2 - 4} \cdot \frac{x - 2}{x}$$

Now factor the denominator of the first term and cancel:

$$x^2 - 4 = (x - 2)(x + 2)$$

$$\frac{1}{(x - 2)(x + 2)} \cdot \frac{x - 2}{x} = \frac{1}{x(x + 2)}$$

30. (4)

Given the lengths of two sides of a triangle, the length of the third side has to be smaller than the sum of the other two sides and larger than their difference:

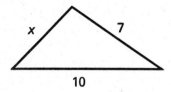

In this case, the length of the third side must be:

$$(10 - 7) < x < (10 + 7)$$
$$3 < x < 17$$

Only answer choice (4) is within this range.

31. (1)

If you know the difference between the formulas for combinations and permutations, you can recognize answer choice (1) as being the different one. The Combinations Formula is: $_nC_r = \dfrac{n!}{r!\,(n-r)!}$

Therefore, the value of $_7C_5$ is:

$$\frac{7!}{5!\,2!} = \frac{7 \times 6 \times 5 \times 4 \times 3 \times 2 \times 1}{5 \times 4 \times 3 \times 2 \times 1 \times (2 \times 1)} = \frac{7 \times 6}{2 \times 1} = \frac{42}{2} = 21$$

The Permutations Formula is:

$$_nP_r = \frac{n!}{(n-r)!}$$

The value of $_7P_5$ is:

$$\frac{7!}{2!} = \frac{7 \times 6 \times 5 \times 4 \times 3 \times 2 \times 1}{2 \times 1} = 7 \times 6 \times 5 \times 4 \times 3 = 1,440$$

Not even close.

If you look at answer choices (2), (3), and (4), you'll find they all equal 21.

32. (3)

The answer choices should tell you that you can't factor this evenly; you have to use the Quadratic Formula:

$$x = \frac{-b \pm \sqrt{b^2 - 4ac}}{2a}$$

In this equation, $a = 2$, $b = -6$, and $c = 3$:

$$x = \frac{-6 \pm \sqrt{(-6)^2 - 4(2)(3)}}{2(2)} = \frac{-6 \pm \sqrt{36 - 24}}{4} = \frac{-6 \pm \sqrt{12}}{4}$$

At this point, you can cross off answer choices (2) and (4), because the right answer won't involve $\sqrt{15}$. It's also clear that the first term in the numerator will be negative, so you can eliminate answer choice (3). Only answer choice (1) is left.

For those of you who want to finish out the math, you first have to reduce $\sqrt{12}$ by factoring out a perfect square:

$$\sqrt{12} = \sqrt{4 \times 3} = \sqrt{4} \times \sqrt{3} = 2\sqrt{3}$$

The answer now becomes:

$$x = \frac{-6 \pm 2\sqrt{3}}{4}$$

Divide each term in the fraction by 2, and your final answer is:

$$x = \frac{-3 \pm \sqrt{3}}{2}$$

33. (3)

You can't do anything until the two fractions have the same denominator. This number is the lowest common denominator (LCD) of x and 4, which is $4x$. To make two fractions compatible, multiply the top and bottom of the first fraction by 4:

$$\frac{(4)}{(4)} \cdot \frac{x+4}{x} = \frac{4x+16}{4x}$$

Multiply the top and bottom of the second fraction by x:

$$\frac{(x)}{(x)} \cdot \frac{x-4}{4} = \frac{x^2 - 4x}{4x}$$

Now you can add the fractions:

$$\frac{4x + 16}{4x} + \frac{x^2 - 4x}{4x} =$$

$$\frac{4x + 16 + x^2 - 4x}{4x} =$$

$$\frac{x^2 + 16}{4x}$$

34. (1)

This sneaky question looks like it involves some complicated probability formula, but it doesn't. It's just a matter of looking at what's possible and what isn't. Look at the triangle again, this time with the median from point B and the altitude drawn in:

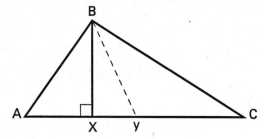

Consider $\triangle BXY$. It's a right triangle with $\overline{BY}$ as its hypotenuse. Since the hypotenuse of a right triangle is always the triangle's longest side, the median will *always* be larger than the altitude. Whenever something always happens, the probability that it will happen is 1. It's an absolute certainty.

35. construction

First, put the pointy end of your compass on point B and make an arc that intersects the two rays of the angle. Label the two points of intersection M and N:

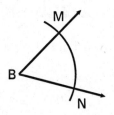

Without changing the width of your compass, place the pointy end on point M and draw an arc inside $\angle B$. Then put the pointy end on point N and draw another arc inside $\angle B$. Mark the point of intersection P·

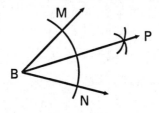

Connect ray BP. That's the angle bisector.

Part II

36. *a*

Factor the algebraic terms in the fractions like this:

$$x^2 - 9 = (x - 3)(x + 3)$$
$$x^2 - x - 20 = (x - 5)(x + 4)$$
$$4x^2 - 20x = 4x(x - 5)$$
$$4x^2 - 12x = 4x(x - 3)$$

Now rewrite the problem like this, and cancel out all the terms that appear both on the top and on the bottom:

$$\frac{(x - 3)(x + 3)}{(x - 5)(x + 4)} \cdot \frac{4x(x - 5)}{4x(x - 3)}$$

$$\frac{(x - 3)(x + 3)}{(x - 5)(x + 4)} \cdot \frac{4x(x - 5)}{4x(x - 3)} = \frac{x + 3}{x + 4}$$

To check your work, you can plug in a value for x.

b −9, 4

Whenever two fractions are equal to each other, you can cross-multiply:

$$\frac{x-3}{2} = \frac{6}{x+8}$$
$$(x-3)(x+8) = 2 \times 6$$
$$x^2 + 5x - 24 = 12$$
$$x^2 + 5x - 36 = 0$$

Now factor the polynomial and set each factor equal to zero:

$$(x+9)(x-4) = 0$$
$$x = \{-9, 4\}$$

Plug these two values back into the equation to make sure they work.

37. ***a***

Figure out the points you have to graph by plugging in all the integers between −1 and 5, inclusive. For example, if $x = -1$, then $y = (-1)^2 - 4(-1) + 4$, or 9. Your first coordinate is (−1, 9). The rest of your T-chart is below, along with the graph of the parabola:

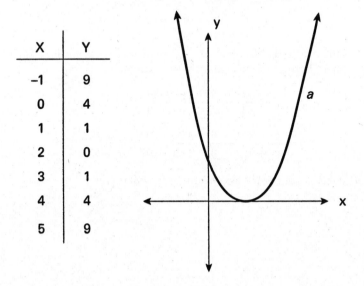

X	Y
−1	9
0	4
1	1
2	0
3	1
4	4
5	9

b

Now, calculate the new points after the translation $(x, y) \rightarrow (x - 2,$ $x + 3)$ by subtracting 2 from each x-coordinate and adding 3 to each y-coordinate:

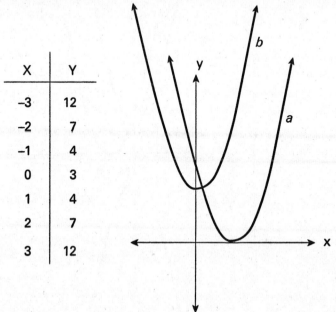

X	Y
–3	12
–2	7
–1	4
0	3
1	4
2	7
3	12

c

After a reflection in the x-axis, each x-coordinate remains the same and each y-coordinate is negated. In other words, $r_{x\text{-axis}}(x, y) \rightarrow (x, -y)$:

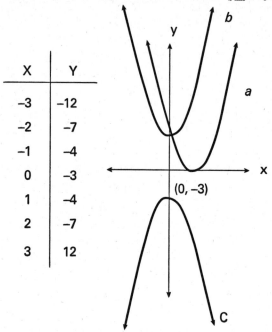

X	Y
-3	-12
-2	-7
-1	-4
0	-3
1	-4
2	-7
3	12

d (3)

Since the parabola from Part C opens downward, the coefficient of x^2 must be *negative*. Therefore, eliminate answer choice (2).

The other key element of this parabola is the y-intercept. When a parabola is expressed in the standard form $y = ax^2 + bx + c$, the c represents the y-intercept. In this case, the parabola intersects the y-axis at the point (0,–3). Therefore, the y-intercept is –3. Of the three remaining answer choices, only (3) has a y-intercept of –3. There's your answer.

Isn't POE wonderful?

38. *a* **120**

This is a combinations problem (because the order in which David chooses the books doesn't matter), so use the Combinations Formula:

$$_nC_r = \frac{n!}{r!\,(n-r)!}$$

There are ten books ($n = 10$), and David chooses three of them ($r = 3$):

$$_{10}C_3 = \frac{10!}{3!\,7!} = \frac{10 \times 9 \times 8 \times 7 \times 6 \times 5 \times 4 \times 3 \times 2 \times 1}{3 \times 2 \times 1 \times (7 \times 6 \times 5 \times 4 \times 3 \times 2 \times 1)}$$

$$= \frac{10 \times 9 \times 8}{3 \times 2 \times 1} = \frac{720}{6} = 120$$

b **(1)** $\dfrac{30}{120}$ **or** $\dfrac{1}{4}$

David can choose one of five comic books, so he has five different choices. For the purposes of this test, this thought can be expressed like this:

$$_5C_1 = 5$$

Using this same logic, there are three possible choices involving detective books ($_3C_1 = 3$) and two possible choices involving books on cars ($_2C_1 = 2$). Multiply these three numbers to determine the number of combinations involving one of each book:

$$5 \times 3 \times 2 = 30$$

From Part A, you know that there are 120 combinations of three books. Therefore, the probability that a combination will have one of each book is $\dfrac{30}{120}$, or $\dfrac{1}{4}$.

(2) $\dfrac{10}{120}$ or $\dfrac{1}{12}$

There are five comic books in the group of ten books, so the probability that David will choose a comic book the first time is $\dfrac{5}{10}$. Now there are nine books left, and four of them are comic books.

The probability that he'll choose a comic book a second time is $\frac{4}{9}$.

After two books are selected, there are eight books left, three of which are comic books. The probability that David will choose a third comic book is $\frac{3}{8}$. To find the aggregate probability that David chooses three comic books, multiply these three separate probabilities together:

$$\frac{5}{10} \cdot \frac{4}{9} \cdot \frac{3}{8} = \frac{60}{720} = \frac{1}{12}$$

(3) 0

This one's easy. Since Alan only has two books about cars, there's no way for him to select only car books in a three-book selection. If an event is impossible, the probability that it will happen is zero.

39. *a* 100

Here's a sketch of the trapezoid:

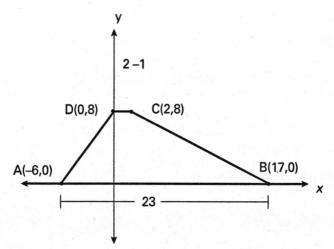

The area of a trapezoid equals $\frac{1}{2}(b_1 + b_2)h$, in which h is the height and b_1 and b_2 are the lengths of the two bases (the two parallel sides).

The top base is 2 units long, and the bottom base is 23 units long (the length from –6 to 17 along the x-axis). The height, as shown below as $\overline{DM}$, is 8 units.

Therefore, the formula becomes:

$$A = \frac{1}{2}(b_1 + b_2)h = \frac{1}{2}(2 + 23)(8) = 100$$

b 52

To find the perimeter of trapezoid $ABCD$, you have to figure out the lengths of $\overline{AD}$ and $\overline{BC}$. Think of each of them as the hypotenuse of a right triangle:

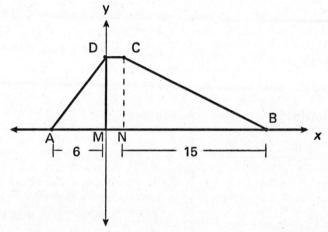

AD is the hypotenuse of $\triangle AMD$, whose legs are AM (6) and DM (8). If you recognize this as a 3:4:5 triangle, you can see right away that AD must equal 10. (Otherwise, use the Pythagorean Theorem.) Similarly, BC is the hypotenuse of $\triangle BNC$, whose legs are BN (8) and NC (15). Use of the Pythagorean Theorem will help you find that $BC = 17$.

Now you can add up all the sides:

$$P = AB + BC + CD + AD = 23 + 17 + 2 + 10 = 52$$

c 28°

Now it's time for a little trigonometry. Look at right $\triangle BNC$. The sine of an angle in a right triangle equals $\dfrac{opposite}{hypotenuse}$. (The SOH in SOHCAHTOA.) Since you know that $BN = 8$ and $BC = 17$, calculate the sine like this:

$$\sin \angle B = \frac{BN}{NC}$$

$$\sin \angle B = \frac{8}{17} = 0.4706$$

Now just push the "$\sin^{-1}$" button on your calculator: $\sin^{-1}(.4706) = 28.07$.

(If you didn't get this, make sure your calculator is in "degree" mode.) They want the answer calculated to the nearest degree, so round off your answer to 28°.

40.

Statements	Reasons
1. $\triangle ABC$, $\overline{AC} \cong \overline{BC}$ $\quad\angle 1 \cong \angle 2$	1. Given
2. CAB CBA	2. If two sides of a triangle are congruent, the angles opposite those two sides are also congruent.
3. $\overline{AB} \cong \overline{BA}$	3. Reflexive Property of Congruence.
4. $\triangle EAB \cong \triangle DBA$	4. ASA $\cong$ ASA
5. $\angle AEB \cong \angle BDA$, $\overline{AE} \cong \overline{BD}$	5. CPCTC
6. $\angle EGA \cong \angle DGB$	6. Vertical angles are congruent.
7. $\triangle EGA \cong \triangle DGB$	7. AAS $\cong$ AAS
8. $\overline{EG} \cong \overline{DG}$	8. CPCTC

Here's the complete proof, with the reasons added:

Statements	Reasons
1. $\triangle ABC$, $\overline{AC} \cong \overline{BC}$, $\angle 1 \cong \angle 2$	1. Given.
2. $\angle CAB \cong \angle CBA$	2. If two sides of a triangle are congruent, the angles opposite those two sides are also congruent.
3. $\overline{AB} \cong \overline{BA}$	3. Reflexive Property of Congruence
4. $\triangle EAB \cong \triangle DBA$	4. ASA $\cong$ ASA
5. $\angle AEB \cong \angle BDA$, $\overline{AE} \cong \overline{BD}$	5. CPCTC
6. $\angle EGA \cong \angle DGB$	6. Vertical angles are congruent.
7. $\triangle EGA \cong \triangle DGB$	7. AAS $\cong$ AAS
8. $\overline{EG} \cong \overline{DG}$	8. CPCTC

Part III

41.

Step One: Turn all the givens into symbolic terms:

"If pro basketball players compete in the Olympics, then college players do not play." $P \rightarrow \sim C$

"If college players do not play, then the team is not an amateur team." $\sim C \rightarrow \sim A$

"If the team is not an amateur team and the team does not win a gold medal, then the people are not happy." $(\sim A \wedge \sim G) \rightarrow \sim H$

"Pro basketball players compete in the Olympics." P

"The people are happy." H

Step Two: Decide what you want to prove:

"The team wins the gold medal." G

Step Three: Write the proof.

Statements	Reasons
1. $P \rightarrow {\sim}C$; ${\sim}C \rightarrow {\sim}A$	1. Given
2. $P \rightarrow {\sim}A$	2. Chain Rule
3. P	3. Given
4. ${\sim}A$	4. Law of Detachment (2, 3)
5. $({\sim}A \wedge {\sim}G) \rightarrow {\sim}H$; H	5. Given
6. ${\sim}({\sim}A \wedge {\sim}G)$	6. Law of *Modus Tollens* (4, 5)
7. $A \vee G$	7. De Morgan's Laws
8. G	8. Law of Disjunctive Inference (4, 7)

42.

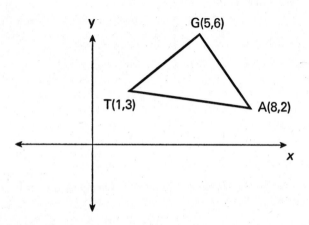

You can prove that $\triangle TAG$ is isosceles by showing that two sides have the same length. Use the distance formula to calculate the length of each side:

$$d = \sqrt{(x_2 - x_1)^2 + (y_2 - y_1)^2}$$

$$AG = \sqrt{(6-2)^2 + (5-8)^2} \qquad TG = \sqrt{(6-3)^2 + (5-1)^2}$$
$$= \sqrt{4^2 + (-3)^2} \qquad\qquad = \sqrt{4^2 + 3^2}$$
$$= \sqrt{16 + 9} \qquad\qquad\quad = \sqrt{16 + 9}$$
$$= \sqrt{25} = 5 \qquad\qquad\quad = \sqrt{25} = 5$$

$$TA = \sqrt{(8-1)^2 + (2-3)^2}$$
$$= \sqrt{7^2 + (-1)^2}$$
$$= \sqrt{49+1}$$
$$= \sqrt{50} = 5\sqrt{2}$$

Since $AG = TG$, $\triangle TAG$ is isosceles.

If the lengths of the three sides fit in the equation of the Pythagorean Theorem, $\triangle TAG$ must also be a right triangle. (Since TA is the largest side, it must be the hypotenuse):

$$(AG)^2 + (TG)^2 = (TA)^2$$
$$5^2 + 5^2 = (5\sqrt{2})^2$$
$$25 + 25 = 50$$

Since $\triangle TAG$ has two equal sides, and the lengths of all three sides work in the Pythagorean Theorem, $\triangle TAG$ must be an isosceles right triangle.

Note: If you noticed that the lengths of the three sides are in a ratio $1{:}1{:}\sqrt{2}$, you could recognize that as the ratio of the sides of a 45:45:90 triangle (which is an isosceles right triangle).

EXAMINATION: AUGUST 1994

Part I

Answer 30 questions from this part. Each correct answer will receive 2 credits. No partial credit will be allowed. Write your answers in the spaces provided on the separate answer sheet. Where applicable, answers may be left in terms of π or in radical form.   [60]

1 In the accompanying diagram, $\overline{DE} \parallel \overline{BC}$, $AD = 8$, $AB = 12$, and $EC = 5$. Find AE.

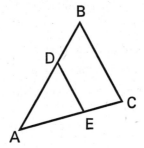

2 If $x * x = x^y + \dfrac{x}{y}$, find the value of $4 * 2$.

3 What are the coordinates of A', the image of point $A(-5,1)$ after a reflection in the y-axis?

4 In a right triangle, the legs have lengths 5 and 7. Express the length of the hypotenuse in radical form.

5 In parallelogram *LMNO*, an exterior angle at vertex *O* measures 72°. Find the measure, in degrees, of ∠*L*.

6 Two parallel lines are cut by a transversal. Two interior angles on the same side of the transversal are represented by $2x$ and $30 + x$. What is the measure of the *smaller* angle?

7 Solve for *a*: $\dfrac{a}{3} + \dfrac{5a}{12} = \dfrac{9}{4}$

8 A club has 12 members. How many different two-person committees can be formed?

9 In $\triangle ABC$, $m\angle A = 41$, $m\angle B = 2x - 37$, and $m\angle C = 3x - 29$. Which side of the triangle is the *shortest* side?

10 In the accompanying diagram of right triangle *ABC*, $\overline{CD}$ is drawn perpendicular to hypotenuse $\overline{AB}$. If $AB = 16$ and $DB = 4$, find *BC*.

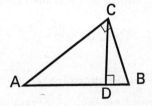

11 A triangle has sides of lengths 12, 14, and 18. Find the perimeter of a similar triangle after a dilation of 2.

12 Solve for the positive value of x: $\dfrac{x-4}{5} = \dfrac{1}{x}$, $x \neq 0$

13 What is the total number of points that are 3 units from line m and also 5 units from P, a point on line m?

14 Factor completely: $x^3 + 5x^2 + 6x$

Directions (15-34): For *each* question chosen, write the *numeral* preceding the word or expression that best completes the statement or answers the question.

15 If a, b, and c are real numbers, which statement is always true?
   (1) $a \div b = b \div a$
   (2) $a(b + c) = (a + b) \times (a + c)$
   (3) $a(b \times c) = (a \times b)c$
   (4) $a \times 0 = a$

16 If the statements $\sim r \rightarrow b$, $b \rightarrow \sim m$, and $\sim r$ are true, which statement must also be true?
   (1) $\sim b$ (3) r
   (2) $\sim m$ (4) m

17 Two triangles are congruent if
   (1) corresponding angles are congruent
   (2) corresponding sides and corresponding angles are congruent
   (3) the angles in each triangle have a sum of 180°
   (4) corresponding sides are proportional

18 In a triangle, one exterior angle measures 36°. What is the probability that the triangle is a right triangle?

(1) 1

(3) $\dfrac{1}{2}$

(2) $\dfrac{2}{3}$

(4) 0

19 If line l is perpendicular to line m and the slope of line l is undefined, what is the slope of line m?

(1) 1

(3) 0

(2) $\dfrac{1}{2}$

(4) –1

20 What is the product of the roots of the equation $x^2 - 2x - 15 = 0$?

(1) –15

(3) –8

(2) –2

(4) 30

21 Which statement is the converse of $a \rightarrow b$?

(1) $a \rightarrow \sim b$

(3) $\sim a \rightarrow \sim b$

(2) $b \rightarrow a$

(4) $\sim b \rightarrow \sim a$

22 What is the equation of the locus of points equidistant from points $A(1,2)$ and $B(5,2)$?

(1) $x = 3$

(3) $x = 2$

(2) $y = 3$

(4) $y = 2$

23 If $\cos x = 0.8$, what is the value of $\sin x$?

(1) 1.0

(3) 0.6

(2) 0.2

(4) 0.4

24 In $\triangle ABC$, $AB = 10$ and $BC = 5$. Which expression can be true?

(1) $AC = 5$ (3) $AC < 5$

(2) $AC = 20$ (4) $AC > 5$

25 How many different eight-letter arrangements can be formed from the word "MONOMIAL"?

(1) $\dfrac{8!}{2!\,2!}$ (3) $8!$

(2) $\dfrac{8!}{2!}$ (4) $6!$

26 Which statement is logically equivalent to $(a \wedge b) \rightarrow c$?

(1) $\sim c \rightarrow (\sim a \wedge \sim b)$

(2) $\sim c \rightarrow (\sim a \vee \sim b)$

(3) $\sim c \rightarrow (\sim a \vee b)$

(4) $\sim c \rightarrow (a \wedge \sim b)$

27 Which is a solution for the following system of equations?

$$y = x^2$$
$$y = -4x + 12$$

(1) $(-2,4)$ (3) $(2,4)$

(2) $(6,36)$ (4) $(-6,24)$

28 Which point is closest to the origin?

(1) $(5,12)$ (3) $(10,4)$

(2) $(6,8)$ (4) $(0,11)$

29 If the endpoints of a diameter of a circle are (2,1) and (4,0), what are the coordinates of the center of the circle?

(1) (6,–1)

(3) $\left(3, \frac{1}{2}\right)$

(2) $\left(3, -\frac{1}{2}\right)$

(4) (2,–1)

30 In the diagram below, m ∠C = 90, m ∠A = 42, and CA = 10.

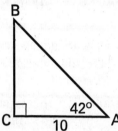

Which equation can be used to find AB?

(1) $\tan 42° = \frac{10}{AB}$

(3) $\cos 42° = \frac{AB}{10}$

(2) $\tan 42° = \frac{AB}{10}$

(4) $\cos 42° = \frac{10}{AB}$

31 The graph of the equation $x - 3y = 6$ is parallel to the graph of

(1) $y = -3x + 7$

(3) $y = 3x - 8$

(2) $y = -\frac{1}{3}x + 5$

(4) $y = \frac{1}{3}x + 8$

32 Which is an equation of the circle with center at (−3,1) and radius of 5?

(1) $(x + 3)^2 + (y − 1)^2 = 5$
(2) $(x − 3)^2 + (y + 1)^2 = 5$
(3) $(x + 3)^2 + (y − 1)^2 = 25$
(4) $(x − 3)^2 + (y + 1)^2 = 25$

33 What are the coordinates of the turning point of the parabola whose equation is $y = x^2 − 2x − 3$?

(1) (1,−4)   (3) (1,2)
(2) (−1,0)   (4) (−1,−2)

34 Let p represent "The diagonals are congruent" and let q represent "The diagonals are perpendicular." For which quadrilateral is $p \wedge \sim q$ true?

(1) parallelogram   (3) square
(2) rhombus   (4) rectangle

Directions (35): Leave all construction lines on the answer sheet.

35 *On the answer sheet*, using a straightedge and compass, locate the midpoint of line segment $\overline{AB}$ and label it *M*.

A •————————————• B

Part II

Answer *three* questions from this part. Clearly indicate the necessary steps, including appropriate formula substitutions, diagrams, graphs, charts, etc. Calculations that may be obtained by mental arithmetic or the calculator do not need to be shown.   [30]

36 For all values of x for which these expressions are defined, perform the indicated operation and express in simplest form.

  a $\dfrac{3x+1}{x^2-1} - \dfrac{1}{x+1}$ [5]

  b $\dfrac{x^2-3x}{x^2+2x} \div \dfrac{x^2-5x+6}{x^2-4}$ [5]

37 Solve the following system of equations and check. [*Either an algebraic or a graphic method will be accepted.*]

$$x^2 + y^2 = 25$$
$$3x - 4y = 0 \quad [8,2]$$

38 In the accompanying diagram of isosceles triangle ABC, $\overline{AB} \cong \overline{CB}$ and altitude BD is 2 more than AD. The area of isosceles triangle ABC is 10.

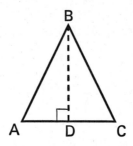

a Find the length of $\overline{AD}$ to the *nearest tenth*. [7]

b Using the answer from part *a*, find the length of $\overline{AB}$ to the *nearest tenth*. [3]

39 *a* On graph paper, draw the graph of the parabola $y = (x + 3) - 2$ for all values of x in the interval $-6 \le x \le 0$. [5]

b On the same set of axes, draw the image of the graph drawn in part *a* after a translation that maps (x,y) to $(x + 3, y + 2)$. [3]

c On the same set of axes, draw the image of the graph drawn in part *b* after a reflection in the x-axis. [2]

40 The operation * is commutative in the table shown below.

*	L	O	G	I	C
L	L	L	O	O	L
O	O			C	O
G		I	G	G	G
I				I	I
C	L	O	G		C

a Copy and complete the table. [2]

b What is the identity element for the operation*? [2]

c What is the inverse of O under the operation*? [2]

d Evaluate: $[O * (G * L) * C$ [2]

e Solve for x: $(C * I) * (G * x) = C$ [2]

Part III

Answer *one* question from this part. Clearly indicate the necessary steps, including appropriate formula substitutions, diagrams, graphs, charts, etc. Calculations that may be obtained by mental arithmetic or the calculator do not need to be shown. [10]

41 Given: quadrilateral ABCD, $\overline{FGE}$, $\overline{AGC}$, $\overline{FG} \cong \overline{EG}$, $\overline{AG} \cong \overline{CG}$, and $\angle B \cong \angle D$.

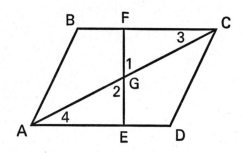

a Prove that $\overline{BC} \cong \overline{DA}$. [7]

b Prove that $ABCD$ is a parallelogram. [3]

42 The coordinates of $\triangle ABC$ are $A(0,0)$, $B(2,6)$, and $C(4,2)$. Using coordinate geometry, prove that if the midpoints of sides $\overline{AB}$ and $\overline{AC}$ are joined, the segment formed is parallel to the third side and equal to one-half the length of the third side. [10]

ANSWER KEY

Part I

1. 10	13. 4	25. (1)
2. 18	14. $x(x + 2)(x + 3)$	26. (1)
3. (5,1)	15. (3)	27. (3)
4. $\sqrt{74}$	16. (2)	28. (2)
5. 72	17. (2)	29. (2)
6. 80	18. (4)	30. (4)
7. 3	19. (3)	31. (4)
8. 66	20. (1)	32. (3)
9. $\overline{BC}$	21. (2)	33. (1)
10. 8	22. (1)	34. (4)
11. 88	23. (3)	35. construction
12. 5	24. (4)	

EXPLANATIONS: AUGUST 1994

Part I

1. 10

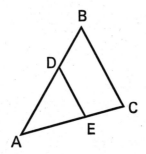

Because $\overline{DE}$ and $\overline{BC}$ are parallel, $\triangle ADE$ and $\triangle ABC$ are similar. Once you establish that, you can find the length of $\overline{AE}$ by setting up a proportion using corresponding lengths:

Since $AC = AE + EC$, and $EC = 5$, you can substitute like this:

Now cross-multiply:

$$8(AE + 5) = 12(AE)$$
$$8(AE) + 40 = 12(AE)$$
$$40 = 4(AE)$$
$$10 = AE$$

Plug 10 back into the equation to check your work.

2. 18

They've already defined the function for you, so all you do is let $x = 4$ and $y = 2$:

$$x * y = x^y + \frac{x}{y}$$

$$4 * 2 = 4^2 + \frac{4}{2} = 16 + 2 = 18$$

3. (5, 1)

Whenever you reflect a point in the y-axis, the x-coordinate is negated and the y-coordinate remains the same. The formula for such a reflection looks like this:

$$r_{y-axis}(x, y) = (-x, y)$$

Therefore, when you reflect the point $A(-5, 1)$ over the y-axis, the coordinates of the resulting point A' are $(5,1)$.

4. $\sqrt{74}$

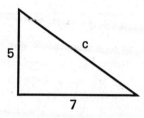

Now use the Pythagorean Theorem. Since the legs are 5 and 7, plug those values in for a and b:

$$c^2 = 5^2 + 7^2$$
$$c^2 = 25 + 49$$
$$c^2 = 74$$
$$c = \sqrt{74}$$

5. 72

The parallelogram looks like this:

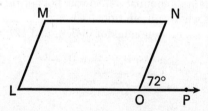

Since the measure of $\angle NOP$ (the external angle at O) measures $72°$, the internal angle ($\angle LON$) measures $108°$ (the two angles are supplementary). Any two consecutive angles of a parallelogram are supplementary, so $\angle L$ must also measure $72°$.

Here's another way to look at it: By definition, sides $\overline{LM}$ and $\overline{NO}$ are parallel. Therefore, $\angle L$ and $\angle NOP$ are corresponding angles. Since corresponding angles are congruent, m $\angle L$ = m $\angle NOP = 72°$

6. 80

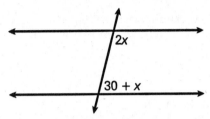

When two lines are cut by a transversal, any two interior angles on the same side of the transversal are supplementary. Therefore:

$$2x + (30 + x) = 180$$
$$3x + 30 = 180$$
$$3x = 150$$
$$x = 50$$

You're not done yet. Now it's time to figure out the size of each angle by plugging in 50 for x:

first angle: $2 \times 50 = 100$; second angle: $30 + 50 = 80$.

The smaller angle measures 80°.

7. 3

Right now, you can't do anything with these fractions because they all have different denominators. So why not get rid of them? Watch what happens when you multiply each term in this equation by 12 (the lowest common denominator of 3, 4, and 12):

$$(12)\frac{a}{3} + (12)\frac{5a}{12} = (12)\frac{9}{4}$$
$$4a + 5a = 27$$
$$9a = 27$$
$$a = 3$$

To check your work, plug 3 back into the equation and make sure it works. The math is a little complicated, but it's worth it.

8. 66

This is a combinations problem (because the order of the members of the committee doesn't matter), so use the Combinations Formula:

$$_nC_r = \frac{n!}{r!\,(n-r)!}$$

$$_{12}C_2 = \frac{12!}{2!\,(10!)} = \frac{12 \times 11 \times 10 \times 9 \times 8 \times 7 \times 6 \times 5 \times 4 \times 3 \times 2 \times 1}{(2 \times 1) \times (10 \times 9 \times 8 \times 7 \times 6 \times 5 \times 4 \times 3 \times 2 \times 1)}$$

$$= \frac{132}{2} = 66$$

9. $\overline{BC}$

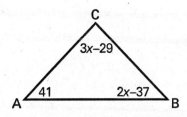

Figure out the size of each triangle using the Rule of 180·

$$\text{m}\angle A + \text{m}\angle B + \text{m}\angle C = 180.$$
$$41 + (2x - 37) + (3x - 29) = 180$$
$$2x + 3x + 41 - 37 - 29 = 180$$
$$5x - 25 = 180$$
$$5x = 205$$
$$x = 41$$

Now find out the measures of $\angle B$ and $\angle C$:

$$\text{m}\angle B = 2(41) - 37 = 45$$
$$\text{m}\angle C = 3(41) - 29 = 94$$

Since $\angle A$ is the *smallest* angle, the side opposite it, $\overline{BC}$, must be the *shortest* side.

10. 8

There are three similar right triangles in the diagram: small (ΔBCD); medium (ΔCAD); and large (ΔABC). For this problem, consider the big one and the small one. Here's how they look if you draw them separate from one another·

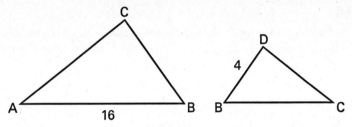

Set up the proportion, involving the short leg and hypotenuse·

$$\frac{BC}{DB} = \frac{AB}{BC}$$

$$\frac{BC}{4} = \frac{16}{BC}$$

Now cross-multiply:

$$BC \times BC = 4 \times 16$$

$$(BC)^2 = 64$$

$$BC = 8$$

Remember that since you're looking for a distance, you don't have to bother with the negative root.

11. **88**

If a triangle undergoes a dilation of 2, each of its sides becomes twice as large:

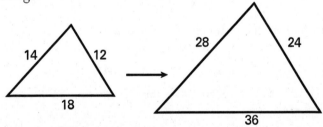

The new triangle has sides of lengths 24, 28, and 36. The perimeter is the sum of all these sides, or 88.

Note: The perimeter of the small triangle is 12 + 14 + 18, or 44. Therefore, it makes sense that a dilation of 2 would double the perimeter to 88.

12. 5

Whenever two fractions are equal to each other, you can cross-multiply:

$$\frac{x-4}{5} = \frac{1}{x}$$
$$x(x-4) = 5 \times 1$$
$$x^2 - 4x = 5$$
$$x^2 - 4x - 5 = 0$$

Now it's just a question of factoring the binomial:

$$(x-5)(x+1) = 0$$
$$x = \{5, -1\}$$

Since they only want the positive root, you can throw out –1.

Now, check your work by plugging 5 back into the equation.

13. 4

Here's a diagram that should help clear this one up:

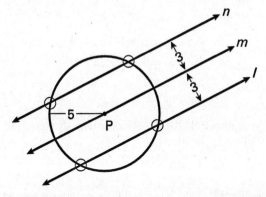

As you can see, the two lines n and p represent all the points that are three units from line m. The circle represents all the points that are five units from point P. There are four points of intersection (A, B, C, and D).

14. $x(x + 2)(x + 3)$

Each term contains an x, so factor it out: $x(x^2 + 5x + 6)$

Now, it's only a matter of factoring the binomial in the parentheses: $(x^2 + 5x + 6) = (x + 3)(x + 2)$

Once you're done, double-check your work using FOIL: $(x + 2)(x + 3) = x^2 + 3x + 2x + 6 = x^2 + 5x + 6$

Multiple Choice

15. **(3)**

Cross off answer choice (4) right away. Any number times zero equals zero. Now try plugging in numbers for the variables. Let's say $a = 2$, $b = 3$, and $c = 5$. Look at answer choice (1) first: $a \div b = \dfrac{2}{3}$ and $b \div a = \dfrac{3}{2}$. That knocks off (1). Similarly, you can eliminate answer choice (2):

$$a(b + c) = (a + b) \times (a + c)$$
$$2(3 + 5) = (2 + 3) \times (2 + 5)$$
$$2(8) = 6 \times 7$$
$$16 \neq 56$$

That leaves answer choice (3), which correctly states the associative property of multiplication.

16. **(2)**

When considered in the proper order, these statements create a perfect syllogism. If $\sim r \to b$ and $b \to \sim m$, you can infer that $\sim r \to \sim m$ by using the Chain Rule (or the law of syllogism). Since you know that r is false ($\sim r$), it must be true that m is also false ($\sim m$) from the Law of Detachment.

17. (2)

If you know the basic properties of similar triangles, you can breeze through this one using POE. Answer choices (1) and (4) can't be right; in similar triangles, corresponding angles are congruent and corresponding sides are proportional. Answer choice (3) is silly; the angles in *every* triangle add up to 180°!

That leaves only answer choice (2), which correctly finishes the statement. If corresponding sides *and* angles of a triangle are congruent, the triangles themselves are congruent.

18. (4)

There's no way that the triangle can be a right triangle. Look at a diagram:

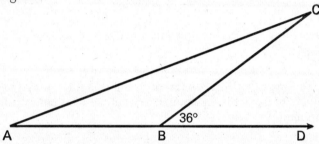

If the external angle measures 36°, then $\angle ABC$ must measure 144°. Using the Rule of 180, you know that the sum of the other two angles must be 36°, so neither of them can be a right angle. Since none of the three angles is a right angle, the triangle is not a right triangle. (No matter how you slice it!)

Whenever something can't happen, the probability that it will happen is zero.

19. (3)

Use POE to solve this one. If two lines are perpendicular, the product of their slopes is –1. Look at each answer choice originally:

(1) If the slope of a line is 1, then a line perpendicular to it has a slope of –1.

(2) If the slope of a line is $\frac{1}{2}$, then a line perpendicular to it has a slope of –2.

(4) If the slope of a line is –1, then a line perpendicular to it has a slope of 1.

None of these slopes is undefined, so you can eliminate each of them.

Note: Here's another way to look at it: The slope of a line is represented by $\frac{\text{rise}}{\text{run}}$. The only way the slope of a line can be undefined is if the denominator (the run) equals zero. Lines with an undefined slope are vertical, since the run (or the difference between the x-coordinates) is zero. If line l is vertical, then line m must be horizontal. Horizontal lines have a slope of zero.

20. (1)

When a binomial is in standard form, the product of its roots is represented by $\frac{c}{a}$. In this equation, $a = 1$, $b = -2$, and $c = -15$, so the product of the roots equals $\frac{-15}{1}$, or –15.

If you forgot that rule, it's just as useful (if a little more time-consuming) to factor the equation, figure out the roots individually, and then multiply them:

$$x^2 - 2x - 15 = 0$$
$$(x - 5)(x + 3) = 0$$
$$x = \{5, -3\}$$

Once you multiply the roots, you see that $-5 \times 3 = -15$.

21. (2)

You can find the converse of a logic statement simply by flipping the letters on either side of the arrow. Thus, the converse of the statement $a \rightarrow b$ is $b \rightarrow a$.

22. (1)

Plot the two points on the coordinate axes like this:

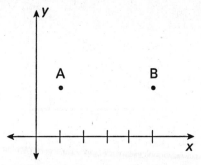

Now, consider the locus of points that are equidistant from the two points. It's a vertical line that looks like this:

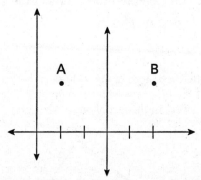

Since the line you're looking for is vertical, you can eliminate answer choices (2) and (4) because they're horizontal. The only vertical line that is equidistant from the points is $x = 3$, because 3 is halfway between 1 and 5.

23. (3)

First, convert 0.8 to a fraction:

$$0.8 = \frac{8}{10} = \frac{4}{5}$$

Now, draw your diagram like this, and remember that cosine involves the CAH in SOHCAHTOA:

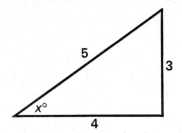

Since $\cos x = \dfrac{4}{5}$, then the leg adjacent to angle x is 4, and the hypotenuse is 5. To find the value of the sine, you have to find the length of the third side, or the side opposite angle x. Since this is a 3:4:5 triangle, the opposite side is 3, and the sine (the SOH in SOHCAHTOA) is $\dfrac{3}{5}$ or 0.6.

24. (4)

In any triangle, the length of any side must be less than the sum of the lengths of the other two sides and greater than their difference. Given the situation in this problem:

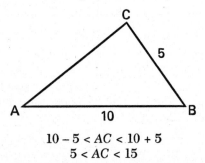

$$10 - 5 < AC < 10 + 5$$
$$5 < AC < 15$$

Given this rule, answer choices (1), (2), and (3) are impossible. Cross 'em out.

25. (1)

If MONOMIAL had no repeated letters, the answer would be answer choice (3).

To find the number of possible arrangements of the letters in a word with n letters, in which one letter appears p times and a different letter appears q times (remember that p and q are greater than 1), the formula looks like this:

$$\frac{n!}{p!\,q!}$$

MONOMIAL has 8 letters, but it has two M's and two O's. Therefore, the number of combinations can be expressed this way:

$$\frac{8!}{2!\,2!}$$

26. (1)

This question requires a lot of logic know-how. The first thing to notice, though, is that each answer choice begins with "~c." That should give you a clue that you're going to use the Law of Contrapositive Inference (the old Flip-and-Negate). The object is to find the negated form of $(a \vee b)$. Using De Morgan's Laws:

$$\sim(a \vee b) \rightarrow (\sim a \wedge \sim b)$$

27. (3)

Problems like this one cry out to be backsolved. Simply plug the values of x and y into the two equations and see which coordinate pair works in both of them. Answer choices (1) and (2) don't fit into $y = -4x + 12$, and answer choice (4) doesn't work for $y = x^2$. Only answer choice (3) works in both of them:

$$y = x^2 \qquad\qquad y = -4x + 12$$
$$4 = (2)^2 \qquad\qquad 4 = -4(2) + 12$$
$$4 = 4 \qquad\qquad\qquad 4 = 4$$

Note: Be sure not to get your coordinates confused: the first number in a coordinate pair is the x-coordinate.

28. (2)

Use geometry to solve this problem. Look at answer choice (1), for example:

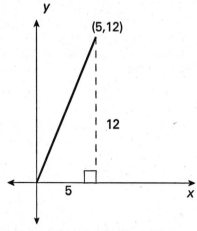

The distance between the origin and point (5,12) is the hypotenuse of a right triangle with legs measuring 5 and 12. You should recognize this as a 5:12:13 triangle, which means the hypotenuse is 13 units long. (If you didn't recognize that, use the Pythagorean Theorem.) Thus, (5,12) is 13 units from the origin.

Using this strategy, you should recognize answer choice (2) as a 6:8:10 triangle (a multiple of a 3:4:5):

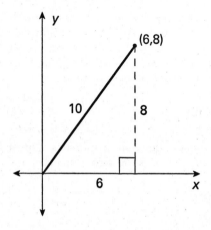

So (6,8) is 10 units away. It's closer than (5,12), so eliminate answer choice (1).

You'll have to use the Pythagorean Theorem for answer choice (3):

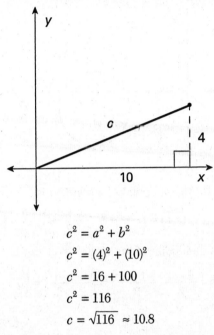

$$c^2 = a^2 + b^2$$
$$c^2 = (4)^2 + (10)^2$$
$$c^2 = 16 + 100$$
$$c^2 = 116$$
$$c = \sqrt{116} \approx 10.8$$

It's still farther away than answer choice (2), so eliminate it.

Answer choice is easiest; since it's on the y-axis, you know it's 11 units from the origin.

Point (6,8) is the closest. Of course, you could always just use the distance formula:

$$d = \sqrt{(x_2 - x_1)^2 + (y_2 - y_1)^2}$$

and calculate the distances that way.

29. (2)

If the endpoints of the diameter of a circle are (2,–1) and (4,0), then the center of the circle must be the midpoint of the segment between those two points:

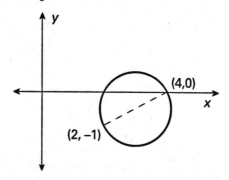

The formula for the midpoint of a segment is:

$$(\bar{x}, \bar{y}) = \left(\frac{x_1 + x_2}{2}, \frac{y_1 + y_2}{2} \right).$$

Therefore, the midpoint of the circle's diameter is:

$$(\bar{x}, \bar{y}) = \left(\frac{2+4}{2}, \frac{-1+0}{2} \right) = \left(\frac{6}{2}, \frac{-1}{2} \right) = \left(3, -\frac{1}{2} \right)$$

30. (4)

$\overline{AB}$ is the hypotenuse of the right triangle, so you'll probably need to use some portion of SOHCAHTOA that involves the hypotenuse. Tangent equals $\frac{\text{opposite}}{\text{adjacent}}$, so it's not going to be much help to you. Eliminate answer choices (1) and (2). From here, it's apparent that you're going to use the cosine (the CAH in SOHCAHTOA). The adjacent leg is 10, and the hypotenuse is unknown. Thus, $\cos 42° = \frac{10}{AB}$.

31. (4)

First, find the slope of the equation in the question by putting it in $y = mx + b$ format:

$$x - 3y = 6$$
$$-3y = -x + 6$$
$$y = \frac{1}{3}x - 2$$

The slope of this line is $\frac{1}{3}$. Since parallel lines all have the same slope, the line in answer choice (4), which also has a slope of $\frac{1}{3}$, is the right answer.

32. (3)

The equation of a circle is the following (remember that (h, k) is the center of the circle and r is its radius):

$$(x - h)^2 + (y - k)^2 = r^2$$

Just plug in the numbers you've been given, and be careful with the minus signs:

$$[x - (-3)]^2 + (y - 1)^2 = 5^2$$
$$(x + 3)^2 + (y - 1)^2 = 25$$

33. (1)

The "turning point" of a parabola is another name for its vertex. You can find the x-coordinate of the vertex using the formula:

$$x = -\frac{b}{2a}$$

In this parabola, $a = 1$, $b = -2$, and $c = -3$:

$$x = -\frac{-2}{2(1)} = 1$$

The x-coordinate of the vertex is 1, so eliminate answer choices (2) and (4). Now, it's just a matter of plugging in 1 for x in the equation and finding the vertex's y-coordinate:

$$y = x^2 - 2x - 3 = (1)^2 - 2(1) - 3 = 1 - 2 - 3 = -4$$

The coordinates of the vertex are $(1, -4)$.

34. (4)

Consider the logical term $p \wedge \sim q$. When you translate it into English, it means "p is true, and q is not true." Now add the statements from the question: "The diagonals are congruent, and the diagonals are NOT perpendicular." Cross off answer choices (1) and (2), because the diagonals of a parallelogram and rhombus don't have to be congruent. Since the diagonals of a square are perpendicular, get rid of answer choice (3). You're left with answer choice (4), which makes sense: the diagonals of a rectangle are congruent, but not necessarily perpendicular.

35.

If you construct the perpendicular bisector of $\overline{AB}$, the point at which the bisector intersects the segment is the segment's midpoint. First, open your compass until it's over half as long as $\overline{AB}$. Then place the pointy end of your compass on A and make two arcs like so:

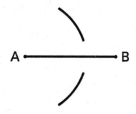

Without changing the width of your compass, put the pointy end on B and find where the two arcs intersect. Label these two points C and D:

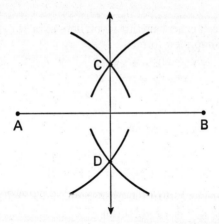

Connect C and D, and find where $\overline{CD}$ intersects $\overline{AB}$. This is point M, the midpoint of $\overline{AB}$.

Part II

36. ***a*** $\quad \dfrac{2}{x-1}$

The key to solving this problem is making the two denominators compatible with each other. Since $x^2 - 1 = (x + 1)(x - 1)$, multiply both the top and bottom of the second term by $(x - 1)$:

$$\frac{1}{x+1} \cdot \frac{(x-1)}{(x-1)} = \frac{x-1}{x^2-1}$$

The denominators are the same, so you can subtract the fractions.

$$\frac{3x+1}{x^2-1} - \frac{x-1}{x^2-1} = \frac{3x+1-(x-1)}{x^2-1} = \frac{2x+2}{x^2-1}$$

Factor $(x + 1)$ out of the top and bottom, and you're done:

$$\frac{2(x+1)}{(x-1)(x+1)} \quad \frac{2}{x-1}$$

b 1

Before you can cancel anything, you have to factor everything down to its most simplified form:

$$x^2 - 3x = x(x - 3)$$
$$x^2 + 2x = x(x + 2)$$
$$x^2 - 5x + 6 = (x - 3)(x - 2)$$
$$x^2 - 4 = (x + 2)(x - 2)$$

Now rewrite the expression:

$$\frac{x(x - 3)}{x(x + 2)} \div \frac{(x - 3)(x - 2)}{(x + 2)(x - 2)}$$

Before you cancel, turn the expression into a multiplication problem by flipping the second term:

$$\frac{x(x - 3)}{x(x + 2)} \times \frac{(x + 2)(x - 2)}{(x - 3)(x - 2)}$$

Everything cancels out, and you're left with 1.

37. **(4, 3), (–4, –3)**

The graphic method might be harder for you if you're artistically challenged. You should recognize the first equation as a circle with a radius of 5 centered at the origin. The other equation is a line. If you put it in $y = mx + b$ format, the equation becomes $y = \frac{3}{4}x$

Thus, the line goes through the origin and has a slope of $\frac{3}{4}$:

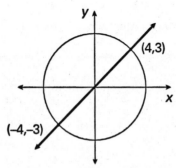

From there, you can see that the two graphs intersect at points (4,3) and (−4,−3). Check your answers by plugging these answers into both equations.

Solving the problem algebraically is a bit more precise. Since $y = \frac{3}{4}x$ (as we saw above), substitute $\frac{3}{4}x$ for y in the first equation:

$$x^2 + y^2 = 25$$

$$x^2 + \left(\frac{3}{4x}\right)^2 = 25$$

$$x^2 + \frac{9}{16}x^2 = 25$$

$$\frac{16}{16}x^2 + \frac{9}{16}x^2 = 25$$

$$\frac{25}{16}x^2 = 25$$

$$\left(\frac{16}{25}\right)\frac{25}{16}x^2 = \left(\frac{16}{25}\right)25$$

$$x^2 = 16$$

$$x = \{-4, 4\}$$

Now, plug −4 and 4 into the second equation:

$3(-4) - 4y = 0$	$3(4) - 4y = 0$
$-12 - 4y = 0$	$12 - 4y = 0$
$-4y = 12$	$-4y = -12$
$y = -3$	$y = 3$

The two points of intersection are (−4,−3) and (4,3).

38. *a* **2.3**

The length you're looking for is AD, so set that equal to x. The altitude, BD, is 2 more than AD, so make BD equal to $x + 2$. Since $\triangle ABC$ is isosceles, the altitude drawn from the vertex angle ($\angle B$) bisects the triangle. Therefore, CD is also equal to x.

Your diagram now looks like this:

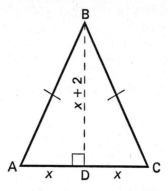

Use the formula for the area of a triangle (the base equals $2x$)·

$$A = \frac{1}{2} bh$$

$$10 = \frac{1}{2}(2x)(x + 2)$$

$$10 = x(x + 2)$$

$$10 = x^2 + 2x$$

$$0 = x^2 + 2x - 10$$

This doesn't factor evenly, so you have to use the Quadratic Formula:

$$x = \frac{-b \pm \sqrt{b^2 - 4ac}}{2a} = \frac{-2 \pm \sqrt{2^2 - 4(1)(-10)}}{2(1)} = \frac{-2 \pm \sqrt{4 + 40}}{2}$$

$$= \frac{-2 \pm \sqrt{44}}{2} = \frac{-2 + 6.6}{2}, \frac{-2 - 6.6}{2} = \frac{4.6}{2}, \frac{-8.6}{2} = 2.3, -4.3$$

Since the answer represents the length of a line segment, throw away the negative result. (You can't have a negative length!) Thus, $AD = 2.3$.

b **4.9**

To find AB, concentrate on the right $\triangle ADB$. From Part A, you know that $AD = 2.3$; therefore, $BD = 2.3 + 2$, or 4.3. Now use the Pythagorean Theorem:

$$(AB)^2 = (2.3)^2 + (4.3)^2$$
$$(AB)^2 = 5.29 + 18.49$$
$$(AB)^2 = 23.78$$
$$AB = 4.9$$

39. a

Start the graphing process by plugging the integers between −6 and 0, inclusive, into the equation for the parabola and determine its coordinates. For example, if $x = -6$, then $y = (-6 + 3)^2 - 2$, or 7. Your first ordered pair is (−6,7). The rest of the coordinates and the graph of the parabola appear below:

x	y
−6	7
−5	2
−4	−1
−3	−2
−2	−1
−1	2
0	7

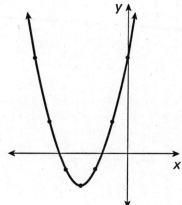

b

To determine the coordinates of the points of the image of the parabola under the translation $(x + 3, y + 2)$, add 3 to each of the x-coordinates and add 2 to each of the y-coordinates. Your sketch should look like this:

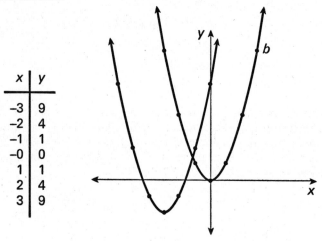

x	y
−3	9
−2	4
−1	1
−0	0
1	1
2	4
3	9

c

Whenever you reflect a point in the x-axis, the x-coordinate remains the same and the y-coordinate is negated. The formula for such a reflection looks like this:

$$r_{x-axis}(x, y) = (x, -y)$$

The coordinates of the new image are below. Your final diagram should look like this:

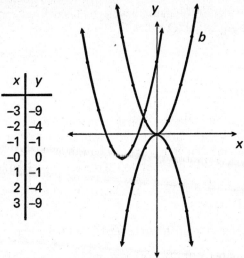

x	y
-3	-9
-2	-4
-1	-1
-0	0
1	-1
2	-4
3	-9

40. a

The finished table looks like this:

*	L	O	G	I	C
L	L	L	O	O	L
O	L	O	I	C	O
G	O	I	G	G	G
I	O	C	G	I	I
C	L	O	G	I	C

b C

Look at each of the rows. At the bottom of the table, you can see that the elements in the row headed by C are the same as the headings of each column:

*	L	O	G	I	C
L	L	L	O	O	L
O	L	O	I	C	O
G	O	I	G	G	G
I	O	C	G	I	I
C	L	O	G	I	C

The identity element is C.

c   I

The inverse of O is the element that, when combined with O, results in the identity element (which you know from Part A to be C). Set up the equation: $(O \circ ?) = C$.

Since $(O \circ I) = C$, it must be true that $? = I$. I must therefore be the inverse of O.

d   O

Solve this problem the way you'd solve any algebra problem involving parentheses and brackets. Start with $(G \circ L)$; find G in the far left column and run your finger along the horizontal row until you hit the column with the L on top. That value is O, so $(G \circ L) = O$.

Now work within the brackets and find the value of $(O \circ O)$ the same way; $(O \circ O) = O$.

You've got one more to go: since $(O \circ C) = O$, that's your final answer.

Here's a summary of the work you did:

$$[O \circ (G \circ L)] \circ C$$
$$[O \circ O] \circ C$$
$$O \circ C$$
$$O$$

e   L

After you solve the first operation, $(C \circ I) = I$, the equation looks like this:

$$I \circ (G \circ x) = C$$

Since $I \circ O = C$, it must be true that $(G \circ x) = O$. Find G in the far left corner and run your finger along that row until you find O. You'll find it in the first column, or the L column. Since $(G \circ L) = O$, $x = L$.

Part III

41. a

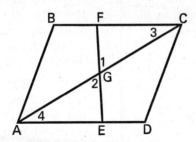

The plan: $\overline{BC}$ and $\overline{DA}$ are corresponding sides of $\triangle ABC$ and $\triangle CDA$ that share a common side, so prove that they're congruent using AAS. First, prove that $\angle 3 \cong \angle 4$ by proving that $\triangle FGC$ and $\triangle AGE$ are congruent.

Statements	Reasons
1. $\overline{FG} \cong \overline{EG}$	1. Given
2. $\angle 1 \cong \angle 2$	2. Vertical angles are congruent.
3. $\overline{AG} \cong \overline{CG}$	3. Given
4. $\triangle FGC \cong \triangle AEG$	4. SAS $\cong$ SAS
5. $\angle 3 \cong \angle 4$	5. CPCTC
6. $\angle B \cong \angle D$	6. Given
7. $\overline{AC} \cong \overline{AC}$	7. Reflexive Property of Equality
8. $\triangle ABC \cong \triangle CDA$	8. AAS $\cong$ AAS
9. $\overline{BC} \cong \overline{DA}$	9. CPCTC

b

The plan: Use the information in Part A, show that $ABCD$ has one pair of sides that are both parallel and congruent, thus proving that it's a parallelogram.

Statements	Reasons
1. $\angle 3 \cong \angle 4$	1. Given (from previous proof)
2. $\overline{BC}$ is parallel to $\overline{AD}$.	2. If two lines can be cut by a transversal such that alternate interior angles are congruent, then the lines are parallel.
3. $\overline{BC} \cong \overline{AD}$	3. Given (from previous proof)
4. $ABCD$ is a parallelogram.	4. If a quadrilateral has two sides that are both parallel and congruent, then the quadrilateral is a parallelogram.

42.

On this problem, you'll use all three major formulas of coordinate geometry: midpoint, distance, and slope.

First, find the midpoints of segments $\overline{AB}$ and $\overline{AC}$ using the midpoint formula:

$$(\bar{x}, \bar{y}) = \left(\frac{x_1 + x_2}{2}, \frac{y_1 + y_2}{2} \right)$$

midpoint of $\overline{AB}$:

$$(\bar{x}, \bar{y}) = \left(\frac{0 + 2}{2}, \frac{0 + 6}{2} \right)$$
$$= \left(\frac{2}{2}, \frac{6}{2} \right)$$
$$= (1, 3)$$

midpoint of $\overline{AC}$:

$$(\bar{x}, \bar{y}) = \left(\frac{0 + 4}{2}, \frac{0 + 2}{2} \right)$$
$$= \left(\frac{4}{2}, \frac{2}{2} \right)$$
$$= (2, 1)$$

Label the two midpoints D and F. Here's what the diagram looks like:

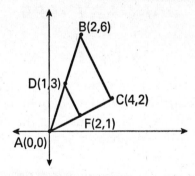

To prove that $\overline{DF}$ and $\overline{BC}$ are parallel, use the slope formula to show that they have the same slope:

$$m = \frac{y_2 - y_1}{x_2 - x_1}$$

Slope of $\overline{DF}$:

$$m = \frac{1-3}{2-1}$$
$$= \frac{-2}{1}$$
$$= -2$$

Slope of $\overline{BC}$

$$m = \frac{2-6}{4-2}$$
$$= \frac{-4}{2}$$
$$= -2$$

The two segments are parallel.

The last step is to show that $\overline{DF}$ is half as long as $\overline{BC}$ using the distance formula.

To find the distance between the two points, use the distance formula:

$$d = \sqrt{(x_2 - x_1)^2 + (y_2 - y_1)^2}$$

$$DF = \sqrt{(2-1)^2 + (1-3)^2}$$
$$= \sqrt{1^2 + (-2)^2}$$
$$= \sqrt{1+4}$$
$$= \sqrt{5}$$

$$BC = \sqrt{(4-2)^2 + (2-6)^2}$$
$$= \sqrt{2^2 + (-4)^2}$$
$$= \sqrt{4+16}$$
$$= \sqrt{20} = 2\sqrt{5}$$

$\overline{DF}$ is half as long as $\overline{BC}$.

EXAMINATION:
JANUARY 1995

Part I

Answer 30 questions from this part. Each correct answer will receive 2 credits. No partial credit will be allowed. Write your answers in the spaces provided on the separate answer sheet. Where applicable, answers may be left in terms of π or in radical form. [60]

1 In the accompanying diagram, $\overleftrightarrow{RS} \parallel \overleftrightarrow{TU}$ and $\overleftrightarrow{GH} \parallel \overleftrightarrow{MN}$. If m $\angle x$ = 115, find m $\angle y$.

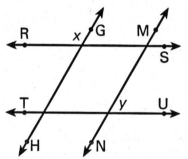

2 In the accompanying diagram, $ABCD$ is a parallelogram with altitude $\overline{DE}$ drawn to side $\overline{AB}$. If $DE = AE$, find the measure of $\angle A$.

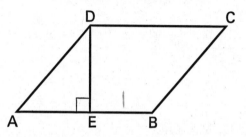

3 The sides of $\triangle ABC$ are 6.8, 6.8, and 8.4 meters. Find the perimeter of the triangle that is formed by joining the midpoints of the sides of $\triangle ABC$.

4 Point $A(6,3)$ is reflected in the x-axis. Find the coordinates of A', its image.

5 If $a \spadesuit b$ is defined as $\dfrac{a-b}{a+b}$, find the value of $-3 \spadesuit 1$.

6 In $\triangle ABC$, side $\overline{AC}$ is extended through C to D and m $\angle DCB = 60$. Which is the longest side of $\triangle ABC$?

7 What is the length of a diagonal of a rectangle whose sides are 3 and 7?

8 Two sides of an isosceles triangle have lengths 2 and 12, respectively. Find the length of the third side.

9 The sides of a triangle have lengths 3, 5, and 7. In a similar triangle, the shortest side has length $x - 3$, and the longest side has length $x + 5$. Find the value of x.

10 Find the number of square units in the area of a triangle whose vertices are $A(2,0)$, $B(6,0)$, and $C(4,5)$.

11 Find, in radical form, the distance between points (−1,−2) and (5,0).

12 What are the coordinates of the center of a circle if the endpoints of a diameter are (−6,2) and (4,6)?

13 In equilateral triangle ABC, $AB = 3x$ and $BC = 2x + 12$. Find the numerical value of the perimeter of $\triangle ABC$.

Directions (17-34): For *each* question chosen, write the *numeral* preceding the word or expression that best completes the statement or answers the question.

14 Which coordinate pair is a solution for the following system of equations?

$$x^2 + y^2 = 8$$
$$x = 2$$

(1) (2,4)   (3) $(2, \sqrt{8})$
(2) (2,2)   (4) (4,2)

15 In the parallelogram $ABCD$, diagonal $\overline{BD}$ is drawn. Which statement must be true?

(1) $\triangle ABD$ must be an obtuse triangle.
(2) $\triangle CDB$ must be an acute triangle.
(3) $\triangle ABD$ must be an isosceles triangle.
(4) $\triangle ABD$ must be congruent to $\triangle CDB$.

16 In the accompanying diagram, AB intersects $\overrightarrow{CD} \perp \overleftrightarrow{AB}$.

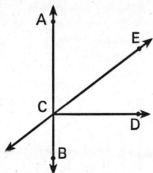

Which statement is true?

(1) $\angle ACE \cong \angle BCD$.
(2) B, C, and D are collinear.
(3) $\angle ACE$ and $\angle ECD$ are complementary.
(4) $\angle ACE$ and $\angle ECD$ are supplementary.

17 Which property is illustrated by
$\square(\Delta + O) = \square \Delta + \square O$?

(1) distributive   (3) commutative
(2) associative   (4) transitive

18 From a deck of 52 cards, two cards are randomly drawn without replacement. What is the probability of drawing two hearts?

(1) $\dfrac{2}{52}$ (3) $\dfrac{13}{52} \cdot \dfrac{12}{51}$

(2) $\dfrac{13}{52} \cdot \dfrac{13}{51}$ (4) $\dfrac{13}{52} \cdot \dfrac{13}{52}$

19 Which is logically equivalent to $\sim(\sim p \vee q)$?

(1) $p \wedge \sim q$ (3) $\sim p \vee \sim q$

(2) $\sim p \wedge \sim q$ (4) $p \vee \sim q$

20 Which is an equation of the circle whose center is the origin and whose radius is 4?

(1) $y = x^2 + 8$ (3) $x^2 + y^2 = 16$

(2) $x^2 + y^2 = 4$ (4) $x + y = 8$

21 Expressed in simplest form, $\frac{x}{2} - \frac{x}{3} + \frac{x}{4}$ is equivalent to

(1) $\frac{x}{3}$ (3) $\frac{3x}{24}$

(2) $\frac{x}{24}$ (4) $\frac{5x}{12}$

22 If a translation maps point $A(-3,1)$ to point $A'(5,5)$, the traslation can be represented by

(1) $(x + 8, y + 4)$ (3) $(x + 2, y + 6)$

(2) $(x + 8, y + 6)$ (4) $(x + 2, y + 4)$

23 When the statement "If A, then B" is true, which statement must also be true?

(1) If B, then A.

(2) If not A, then B.

(3) If not B, then A.

(4) If not B, then not A.

24 In right triangle ABC, altitude $\overline{CD}$ is drawn to hypotenuse $\overline{AB}$. If $AD = 5$ and $DB = 24$, what is the length of $\overline{CD}$?

(1) 120   (3) $2\sqrt{30}$

(2) $\sqrt{30}$ (4) $4\sqrt{30}$

25 The graph of which equation has a *negative* slope?

(1) $y = 5x - 3$ (3) $y = -2 = 4x$

(2) $x + y = 5$ (4) $y = 0$

26 What is the equation of the locus of points passing through point (3,2) and 3 units from the y-axis?

(1) $x = -2$ (3) $x = 3$

(3) $y = -2$ (4) $y = 3$

27 Which expression is a perfect square?

(1) $x^2 - 4x + 4$ (3) $x^2 - 9x + 9$

(2) $x^2 - 4x - 4$ (4) $x^2 - 9x - 9$

28 The roots of the equation $2x^2 + 5x - 2 = 0$ are

(1) $\dfrac{5 \pm \sqrt{41}}{2}$ (3) $2, \dfrac{1}{2}$

(2) $-\dfrac{1}{2}, -2$ (4) $\dfrac{-5 \pm \sqrt{41}}{4}$

29 How many different 13-letter permutations can be formed from the letters of the word "QUADRILATERAL"?

(1) $13!$

(3) $\dfrac{13!}{3!\,2!\,2!}$

(2) $\dfrac{13!}{7!}$

(4) $\dfrac{13!}{6!}$

30 The hypotenuse of right triangle ABC is 10 and $m\angle A = 60$. What is the measure, to the *nearest tenth*, of the leg opposite $\angle A$?

(1) 5.0

(3) 7.1

(2) 5.8

(4) 8.7

31 Which equation represents a line parallel to the line whose equation is $y = 2x - 7$?

(1) $y = 2x$

(3) $y = -7$

(2) $y = \dfrac{1}{2}x - 7$

(4) $y = -\dfrac{1}{2}x + 7$

32 Expressed in simplest form, $\dfrac{2x^2}{x^2 - 1} \cdot \dfrac{x - 1}{x}$, $x \neq 1, 0, -1$, is equivalent to

(1) $\dfrac{2x}{x - 1}$

(3) $\dfrac{2}{x}$

(2) 2

(4) $\dfrac{2x}{x + 1}$

33 A set contains five isosceles trapezoids, three squares, and a rhombus that is not a square. A figure is chosen at random. What is the probability that its diagonals will be congruent?

(1) 1

(3) $\frac{5}{9}$

(2) $\frac{8}{9}$

(4) $\frac{3}{9}$

34 Which graph could represent the equation $y = x^2 - 4$?

(1)

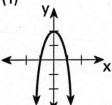

(3)

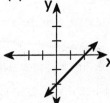

(2)

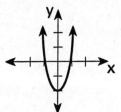

(4)

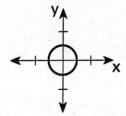

35 In the accompanying diagram, $\triangle ABC$ is scalene.

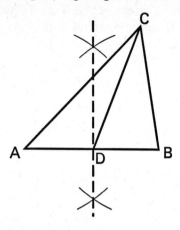

The construction of this triangle shows that $\overline{CD}$ is the

(1) median to side $\overline{AB}$

(2) bisector of angle C

(3) altitude to side $\overline{AB}$

(4) perpendicular bisector to side $\overline{AB}$

Part II

Answer *three* questions from this part. Clearly indicate the necessary steps, including appropriate formula substitutions, diagrams, graphs, charts, etc. Calculations that may be obtained by mental arithmetic or the calculator do not need to be shown. [30]

36 Answer both *a* and *b* for all values of *x* for which these expressions are defined.

 a Solve for x: $\dfrac{x + 3}{3x} = \dfrac{x}{12}$ [5]

 b Express the product as a single fraction in lowests terms:

$$\frac{x}{3x + 15} \bullet \frac{2x^2 + 11x + 5}{2x^2 + x} \quad \text{[5]}$$

37 *a* On graph paper, draw the graph of the equation $y = x^2 + 4x - 1$, including all values of x in the interval $-1 \le x \le 5$. [5]

 b On the same set of axes, graw the graph of the equation $x - y = 5$. [3]

 c From the graphs drawn in part *a* and *b*, determine the solution(s) of this system of equations:
$$y = -x^2 + 4x - 1$$
$$x - y = 5 \quad \text{[2]}$$

38 There are seven boys and three girls on a school tennis team. The coach must select four people from this group to participate in a county championship.

a How many four-person teams can be formed from the group of ten students? [3]

b In how many ways can two boys and two girls be chosen to participate in the county championship? [3]

c What is the probability that two boys and two girls are chosen for the team? [2]

d What is the probability that a four-member team will contain at least one boy? [2]

39 In the accompanying diagram of rhombus *ABCD*, m $\angle BAD$ = 36 and the length of diagonal $\overline{AEC}$ = 16.

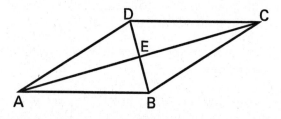

a Find the length of diagonal $\overline{BD}$ to the *nearest tenth*. [4]

b Find the perimeter of rhombus *ABCD* to the *nearest integer*. [6]

40 The endpoints of $\overline{AB}$ are A(1,4) and B(5,1).

  a On graph paper, draw and label $\overline{AB}$. [1]

  b Graph and state the coordinates of $\overline{A'B'}$, the image of $\overline{AB}$ under a reflection in the y-axis. [2]

  c Graph and state the coordinates of $\overline{A''B''}$, the image of $\overline{AB}$ under a dilation of 2 with respect to the origin. [2]

  d Using coordinate geometry, show that a line segment and its image are congruent under a line reflection and are *not* congruent under a dilation. [5]

Part III

Answer *one* question from this part. Clearly indicate the necessary steps, including appropriate formula substitutions, diagrams, graphs, charts, etc. Calculations that may be obtained by mental arithmetic or the calculator do not need to be shown. [10]

41 Given: $B \rightarrow D$

$D \rightarrow \sim E$

$(\sim A \land \sim B) \rightarrow C$

$\sim F \rightarrow E$

$\sim C$

Prove: F [10]

42 Prove: In an isosceles triangle, the line segment that bisects the vertex angle bisects the base. [10]

ANSWER KEY

Part I

1. 65
2. 45
3. 11
4. (6,−3)
5. 2
6. $\overline{AB}$
7. $\sqrt{58}$
8. 12
9. 9
10. 10
11. $\sqrt{40}$ or $2\sqrt{10}$
12. (−1,4)

13. 108
14. (2)
15. (4)
16. (3)
17. (1)
18. (3)
19. (1)
20. (3)
21. (4)
22. (1)
23. (4)
24. (3)

25. (2)
26. (3)
27. (1)
28. (4)
29. (3)
30. (4)
31. (1)
32. (4)
33. (2)
34. (2)
35. (1)

EXPLANATIONS: JANUARY 1995

Part I

1. 65

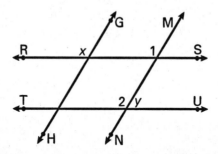

Since line *HG* is parallel to line *MN*, ∠*x* and ∠1 are corresponding angles. Therefore, they have the same measure. Similarly, line *RS* is parallel to line *TU*, so ∠1 and ∠2 are corresponding angles that also have the same measure. Since m∠*x* = m∠1 and m∠1 = m∠2, it must be true that m∠*x* = m∠2. Therefore, m∠2 = 115°.

Since ∠2 and ∠*y* are supplementary, their sum must be 180°:

$$m\angle 2 + m\angle y = 180$$
$$115 + m\angle y = 180$$
$$m\angle y = 65$$

2. 45

Since *DE* = *AE*, △*AED* is isosceles and m∠*DAE* = m∠*EDA*. (If two sides of a triangle are equal in length, then the angles opposite them are equal in measure.)

Further, △*AED* is a right angle. Using the Rule of 180:

$$m\angle DAE + m\angle EDA + m\angle AED = 180$$
$$m\angle DAE + m\angle EDA + 90 = 180$$
$$m\angle DAE + m\angle EDA = 90$$

Since ∠*DAE* and ∠*EDA* have the same measure, they must both measure 45°.

3. 11

Draw your triangle like this:

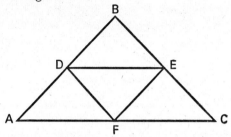

Look first at segment $\overline{DE}$. A segment that connects the midpoints of two sides of a triangle is half the length of the third side. Since E is the midpoint of $\overline{BC}$ and D is the midpoint of $\overline{AB}$, $DE = 4.2$ meters. Similarly, $DF = 3.4$ and $EF = 3.4$. Add these lengths up, and you'll have the perimeter of smaller triangle DEF:

$$4.2 + 3.4 + 3.4 = 11.0 \text{ meters}$$

4. (6, –3)

After a reflection in the x-axis, each x-coordinate remains the same and each y-coordinate is negated. In other words, $r_{x\text{-axis}}(x, y) \rightarrow (x, -y)$. Therefore, when point $A(6,3)$ is reflected in the x-axis, the coordinates of the image of point A' are $(6,-3)$.

5. 2

They've defined what the ♠ means, so all you have to do is plug in $a = -3$ and $b = 1$:

$$a \spadesuit b = \frac{a-b}{a+b} = -3 \spadesuit 1 = \frac{-3-1}{-3+1} = \frac{-4}{-2} = 2$$

6. $\overline{AB}$

Draw your diagram first:

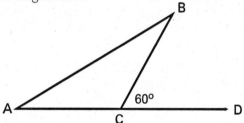

Since m $\angle DCB = 60$, you can deduce that m $\angle ACB = 120$ (because the two angles are supplementary). The sum of the measures of the three angles in a triangle is 180°, so m $\angle A$ + m $\angle B = 60$. This means that $\angle ACB$ must be the biggest angle in the triangle. The side opposite $\angle ACB$ must be the longest side. This side is $\overline{AB}$.

7. $\sqrt{58}$

Think of the rectangle's diagonal as the hypotenuse of a right triangle:

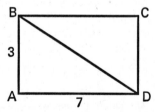

Now use the Pythagorean Theorem to find the length of hypotenuse $\overline{BD}$:

$$(BD)^2 = (AB)^2 + (AD)^2$$
$$(BD)^2 = 3^2 + 7^2$$
$$(BD)^2 = 9 + 49$$
$$(BD)^2 = 58$$
$$BD = \sqrt{58}$$

Since 58 contains no factors that are perfect squares, you can't reduce this radical any further.

8. 12

Since the triangle is isosceles, the third side must have a length of 2 or 12. Given the lengths of two sides of a triangle, the length of the third side has to be smaller than the sum of the other two sides and larger than their difference:

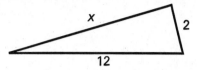

In this case, the length of the third side must be:

$$(12 - 2) < x < (12 + 2)$$
$$10 < x < 14$$

It's not possible for the third side to have a length of 2, because 2 doesn't fall within that range. Therefore, the third side must have a length of 12.

9. 9

Draw your diagram first, and remember which sides correspond with each other:

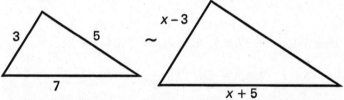

The side of length $(x - 3)$ lines up with the side of length 3 (they're the shortest sides), and the side of length $(x + 5)$ corresponds with the side of length 7 (they're the longest sides). Just set up the proportion:

$$\frac{x - 3}{x + 5} = \frac{3}{7}$$

Now cross-multiply:

$$7(x - 3) = 3(x + 5)$$
$$7x - 21 = 3x + 15$$
$$4x - 21 = 15$$
$$4x = 36$$
$$x = 9$$

To check your work, plug 9 back into the original equation and make sure it works.

10. 10

Plot your points on the coordinate axes, like this:

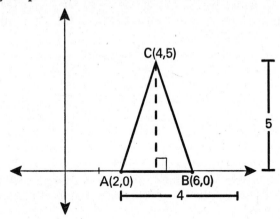

The base of the triangle, side $\overline{AB}$, is 4 units long. The height of the triangle, which is the perpendicular distance from point C to the base, is 5 units. The formula for the area of a triangle is:

$$A = \frac{1}{2} bh$$

Plug in $b = 4$ and $h = 5$, and you're off and running:

$$A = \frac{1}{2}(4)(5) = \frac{1}{2}(20) = 10$$

11. $\sqrt{40}$ or $2\sqrt{10}$

To find the distance between two points on the coordinate axes, use the distance formula:

$$d = \sqrt{(x_2 - x_1)^2 + (y_2 - y_1)^2}$$
$$= \sqrt{[0 - (-2)]^2 + [5 - (-1)]^2}$$
$$= \sqrt{2^2 + 6^2} = \sqrt{4 + 36} = \sqrt{40}$$

To reduce this radical, factor out a perfect square:

$$\sqrt{40} = \sqrt{4 \times 10} = \sqrt{4} \times \sqrt{10} = 2\sqrt{10}$$

Either answer is acceptable.

12. (–1, 4)

If the endpoints of the diameter of a circle are (–6,2) and (4,6), then the center of the circle must be the midpoint of the segment between those two points:

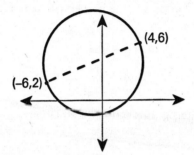

The formula for the midpoint of a segment is:

$$(\bar{x}, \bar{y}) = \left(\frac{x_1 + x_2}{2}, \frac{y_1 + y_2}{2} \right)$$

Therefore, the midpoint of the circle's diameter is:

$$(\bar{x}, \bar{y}) = \left(\frac{-6 + 4}{2}, \frac{2 + 6}{2} \right) = \left(\frac{-2}{2}, \frac{8}{2} \right) = (-1,4)$$

13. 108

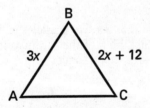

The sides of an equilateral triangle are all equal in length. Therefore, $AB = BC$. Set up an equation to determine the value of x:

$$3x = 2x + 12$$
$$x = 12$$

The length of AB is 3 × 12, or 36. (To test your work further, $BC = (2 \times 12) + 12$, or 36.) The length of each side of the equilateral triangle is 36, so the perimeter of the triangle is 36 + 36 + 36, or 108.

Multiple Choice

14. (2)

At first glance, you can eliminate answer choice (4), because $x = 2$.
Now just plug 2 in for x in the first equation and find the value of y:

$$2^2 + y^2 = 8$$
$$4 + y^2 = 8$$
$$y^2 = 4$$
$$y = \{-2, 2\}$$

The two solutions to this system of equations are (2, 2) and (2, –2),
so answer choice (2) is correct.

15. (4)

You can use POE to solve this one. Look at these two diagrams:

When you draw the diagonal of a parallelogram, you can't predict
the shape of the two triangles you create. The triangles on the left
are acute, and the ones on the right are obtuse. And none of the
triangles has to be isosceles.

It is true, however, that the two triangles will be congruent to each
other. To prove this, think of the formulas for the area of a parallelogram and the area of a triangle:

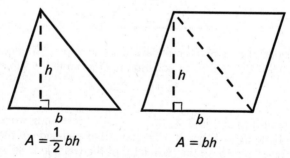

$$A = \frac{1}{2}bh \qquad\qquad A = bh$$

The area of the triangle is half that of the parallelogram.

16. (3)

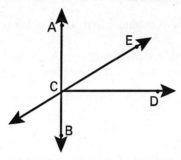

The key bit of information here is that $\overline{CD} \perp \overline{AB}$. That means m$\angle ACD$ = 90. From the Angle Addition Postulate, m$\angle ACE$ + m$\angle ECD$ = m$\angle ACD$. Therefore, the sum of $\angle ACE$ and $\angle ECD$ must be 90°; the two angles are *complementary*.

Many students confuse the terms "complementary" and "supplementary" (which means that the sum of the two angles is 180°), and the Regents people know this.

17. (1)

This problem is a funky way of testing your knowledge of a few basic properties of mathematics. The equation in the problem is an example of the distributive property, which states:

$$a(b + c) = ab + ac$$

They've just substituted shapes for the variables.

Here's a quick review of the other properties in the wrong answers:

(2) associative: $a + (b + c) = (a + b) + c$.

(3) commutative: $a + b = b + a; ab = ba$.

(4) transitive: if $a = b$ and $b = c$, then $a = c$.

18. (3)

There are 13 hearts in a full deck of 52 cards. Therefore, the probability that you'll pick a heart at random is $\dfrac{13}{52}$. The key phrase here is "without replacement." One you've picked a heart, there are only

12 left. Furthermore, you took a card from the deck, so there are only 51 left. The probability that you'll pick a second heart is $\frac{12}{51}$.

To find the combined probability of two events happening consecutively, you have to multiply the individual rates together. Therefore, the probability of drawing two hearts in a row is $\frac{13}{52} \times \frac{12}{51}$.

19. (1)

Whenever you see the negation of a logic statement in parentheses, think of De Morgan's Laws.

$$\sim(a \wedge b) \rightarrow \sim a \vee \sim b$$

This basically means than when you negate a parenthetical statement with a "∧" or "∨" in it, negate each symbol and turn the symbol upside down:

$$\sim(\sim p \vee q) \rightarrow \sim(\sim p) \wedge \sim q$$

Since $\sim(\sim p)$ is the same thing as p (because of the rule of double negation), you can rewrite the statement as: $p \wedge \sim q$.

20. (3)

The equation of a circle is the following (remember that (h, k) is the center of the circle and r is its radius):

$$(x - h)^2 + (y - k)^2 = r^2$$

Just plug in the numbers you've been given. Since the circle is centered at the origin, both h and k are equal to 0:

$$(x - 0)^2 + (y - 0)^2 = 4^2$$
$$x^2 + y^2 = 16$$

An equation can't represent a circle unless it contains both an x^2 term and a y^2 term. Therefore, you can eliminate answer choice (1), which is a parabola, and answer choice (4), which is a line.

21. (4)

You can't do anything until the three fractions have the same denominator, which is the lowest common denominator (LCD) of 2, 3, and 4. The LCD is 12. To convert the first fraction, multiply both the top and bottom by 6:

$$\frac{x}{2} \cdot \frac{6}{6} = \frac{6x}{12}$$

To convert the second fraction, multiply both the top and bottom by 4:

$$\frac{x}{3} \cdot \frac{4}{4} = \frac{4x}{12}$$

To convert the third fraction, multiply both the top and bottom by 3:

$$\frac{x}{4} \cdot \frac{3}{3} = \frac{3x}{12}$$

Now you can combine the numerators:

$$\frac{6x}{12} - \frac{4x}{12} + \frac{3x}{12} = \frac{5x}{12}$$

22. (1)

If a translation maps the point $A(-3,1)$ to point $A'(5,5)$, that translation must involve adding 8 to the x-coordinate (because $-3 + 8 = 5$) and adding 4 to the y-coordinate (because $1 + 4 = 5$). The translation can therefore be written as $(x + 8, y + 4)$.

23. (4)

Hopefully, you recognized answer choice (4) as the contrapositive of the statement in the question. The statement "If A, then B" can be written symbolically like this:

$$A \rightarrow B$$

Using the Flip-and-Negate Rule (otherwise known as the Law of Contrapositive Inference), you can rewrite the statement like this:

$$\sim B \rightarrow \sim A$$

This means "If not B, then not A."

24. (3)

Draw the diagram first:

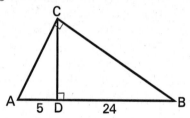

This figure represents three similar right triangles ($\triangle ADC$, $\triangle CDB$, and $\triangle ACB$), and all of their corresponding sides are proportional to each other. For this problem, consider $\triangle ADC$ and $\triangle CDB$ and set up a proportion. In $\triangle CDB$, DB is the long leg and CD is the short leg; in $\triangle ADC$, CD is the long leg and AD is the short leg. Therefore:

$$\frac{DB}{CD} = \frac{CD}{AD}$$

$$\frac{24}{CD} = \frac{CD}{5}$$

When you cross-multiply, you'll get:

$$(CD)^2 = 120$$

$$CD = \pm\sqrt{120}$$

Since you're looking for a distance, you can forget about the negative root. However, you still have a bit more work to do; you have to reduce the square root like this:

$$\sqrt{120} = \sqrt{4 \times 30} = \sqrt{4} \times \sqrt{30} = 2\sqrt{30}$$

25. (2)

If you put each of the four lines into $y = mx + b$ format, you can figure out the slope of each one right away. (The m in the formula represents the slope.)

Answer choice (1) is already in the proper format, and its slope is 5. Eliminate it.

If you subtract x from both sides of the equation in answer choice (2), the result is: $y = -x + 5$. The slope of this line is -1. You don't have to go any further.

Answer choice (3) becomes $y = 4x + 2$, so its slope is 4. As for answer choice (4), the line $y = 0$ coincides with the x-axis, which is horizontal and has a slope of 0.

26. (3)

The locus of points in the question looks like this:

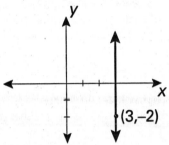

Any locus of points a certain distance from the y-axis, which is vertical, will also be vertical. Hence, you can eliminate answer choices (2) and (4), which are horizontal. Since the x-coordinate of $(3, -2)$ is 3, you know that the x-coordinate of *every* point on the line in the diagram is equal to 3. Thus, the equation of the line must be $x = 3$.

27. (1)

This question is a test of your factoring skills. Since $x^2 - 4x + 4$ can be factored into $(x - 2)(x - 2)$, or $(x - 2)^2$, answer choice (1) is a perfect square.

You can't factor any of the other answer choices. (Be careful about answer choice (3); $(x - 3)^2 = x^2 - 6x + 9$.)

28. (4)

No matter how hard you try, you can't factor the equation. (Cross off answer choices (2) and (3), which assume you can.) You have to use the Quadratic Formula:

$$x = \frac{-b \pm \sqrt{b^2 - 4ac}}{2a}$$

In the equation $2x^2 + 5x - 2 = 0$, $a = 2$, $b = 5$, and $c = -2$:

$$x = \frac{-5 \pm \sqrt{(5)^2 - 4(2)(-2)}}{2(2)} = \frac{-5 \pm \sqrt{25 + 16}}{2} = \frac{-5 \pm \sqrt{41}}{2}$$

29. (3)

Answer choice (1) would be correct if there were no repeating letters. Since we have some repetition, however, eliminate it.

The formula to follow is a variation of the permutations rule. To find the number of possible arrangements of the letters in a word with n letters, in which one letter appears p times, another letter appears q times, and a third letter appears r times (remember that p, q, and r are all greater than 1), the formula looks like this:

$$\frac{n!}{p!q!r!}$$

QUADRILATERAL has 13 letters, but there are three A's, two R's, and two L's. Therefore, you can express the number of arrangements as:

$$\frac{13!}{3!2!2!}$$

30. (4)

Create your diagram like this, and make sure to label $\angle A$ correctly:

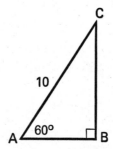

Since you know the hypotenuse of the triangle (10) and you're looking for the leg *opposite* $\angle A$, you need to use the trig function that involves opposite and hypotenuse—the sine (the SOH in SOHCAHTOA). Set up your equation like this:

$$\sin \angle A = \frac{\text{opposite}}{\text{hypotenuse}}$$

$$\sin 60° = \frac{BC}{10}$$

$$0.8660 = \frac{BC}{10}$$

$$10(0.8660) = BC$$

$$8.66 = BC$$

When you round your answer off to the nearest *tenth* (as instructed), you get 8.7.

31. (1)

Parallel lines have the same slope. Since the line in the question, *y* = 2*x* − 7, is in standard *y* = *mx* + *b* format, you know that its slope (the *m*) is 2. The line in answer choice (1) also has a slope of 2, so it's parallel to the line in the question.

32. (4)

There's only one algebraic term to factor: $x^2 - 1 = (x - 1)(x + 1)$

Now rewrite the problem like this, and cancel out all the terms that appear both on the top and on the bottom:

$$\frac{2x^2}{(x - 1)(x + 1)} \cdot \frac{x - 1}{x}$$

$$\frac{2x^2}{(x - 1)(x + 1)} \times \frac{x - 1}{x} = \frac{2x}{x + 1}$$

33. (2)

Of the three types of quadrilaterals mentioned in the problem, two of them have congruent diagonals: the isosceles trapezoids and the squares (see below).

The only quadrilateral that doesn't have congruent diagonals is the rhombus that is not a square:

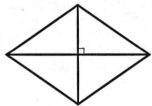

Therefore, there are eight quadrilaterals that have congruent diagonals out of a possible nine. The probability of selecting a quadrilateral with congruent diagonals is $\frac{8}{9}$.

34. (2)

In the equation $y = x^2 - 4$, the x is squared, but the y is not. You should recognize this as a parabola. Eliminate answer choices (3) and (4).

The equation is in standard form $y = ax^2 + bx + c$, and $a = 1$. Since a is positive, the parabola must open up. (Remember: if a is positive, the parabola smiles. If a is negative, the parabola frowns.) The answer must be answer choice (2).

Note: Bear in mind that the y-intercept of the parabola is -4, which also makes answer choice (2) the only possible correct answer.

35. (1)

In the diagram, the perpendicular bisector of $\overline{AB}$ has been constructed; D is the midpoint of $\overline{AB}$. The line segment that connects a vertex of a triangle to the midpoint of the opposite side of that triangle is called the median, so answer choice (1) is correct.

Part II

36. *a* $x = \{6, -2\}$

Whenever two fractions are equal to each other, you can cross-multiply:

$$\frac{x + 3}{3x} = \frac{x}{12}$$
$$12(x + 3) = 3x(x)$$
$$12x + 36 = 3x^2$$
$$-3x^2 + 12x + 36 = 0$$

To make this equation a little easier to factor, divide each term by –3:

$$x^2 - 4x - 12 = 0$$

Now factor the polynomial and set each factor equal to zero:

$$(x - 6)(x + 2) = 0$$
$$x = \{6, -2\}$$

Plug these two values back into the equation to make sure they work.

b $\dfrac{1}{3}$

It's hard to believe that something so complicated can reduce to something so simple, but it's possible. First, factor all the complex terms
like this:

$$2x^2 + 11x + 5 = (2x + 1)(x + 5)$$
$$3x + 15 = 3(x + 5)$$
$$2x^2 + x = x(2x + 1)$$

Once you've factored these four terms, the problem looks like this:

$$\frac{x}{3(x + 5)} \cdot \frac{(2x + 1)(x + 5)}{x(2x + 1)}$$

Cancel out all the factors that appear both on the top and on the
bottom, and you're left with:

$$\frac{x}{3\cancel{(x + 5)}} \cdot \frac{\cancel{(2x + 1)}\cancel{(x + 5)}}{\cancel{x}\cancel{(2x + 1)}} = \frac{1}{3}$$

37. *a*

Figure out the points you have to graph by plugging in all the integers between –1 and 5, inclusive. For example, if $x = -1$. Then $y = -(-1)^2 + 4(-1) -1$, or –6. Your first ordered pair is $(-1, -6)$. The rest of the T-chart and the graph appear below:

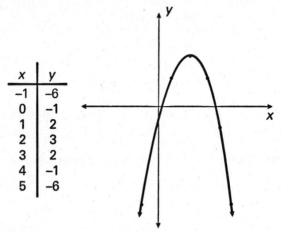

x	y
–1	–6
0	–1
1	2
2	3
3	2
4	–1
5	–6

b

This graph is a little easier; it's only a line that looks like this:

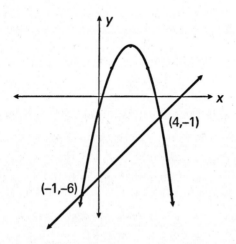

c **(–1, –6) and (4, –1)**

Your ability to find the two points of intersection relies on how accurately you drew the previous two diagrams. They want you to use the graphs, so estimate the points and check them by plugging them into the equations:

$$y = -x^2 + 4x - 1$$

$-6 = -(-1)^2 + 4(-1) - 1$	$-1 = -4^2 + 4(4) - 1$
$-6 = -1 - 4 - 1$	$-1 = -16 + 16 - 1$
$-6 = -6$ Check.	$-1 = -1$ Check.

If both points work in both equations, you know you got it right.

38. *a* 210

This problem promises to be full of work with combinations, because the order of the tennis players doesn't matter. If you're finding the number of combinations of r people that you can choose from a group of n people, the formula is:

$$_nC_r = \frac{n!}{r!(n-r)!}$$

To find the number of ways you can choose four tennis players from a group of 10, you need to find the value of $_{10}C_4$, which equals:

$$\frac{10!}{4!6!} = \frac{10 \times 9 \times 8 \times 7 \times 6 \times 5 \times 4 \times 3 \times 2 \times 1}{4 \times 3 \times 2 \times 1 \times (6 \times 5 \times 4 \times 3 \times 2 \times 1)}$$

$$= \frac{10 \times 9 \times 8 \times 7}{4 \times 3 \times 2 \times 1} = \frac{630}{3} = 210$$

***b* 63**

This one's a bit more complicated. You want to find the number of ways 2 out of 7 boys can be chosen ($_7C_2$) and the number of ways 2 out of 3 girls can be chosen ($_3C_2$):

$\dfrac{7!}{2!5!} = \dfrac{7 \times 6 \times 5 \times 4 \times 3 \times 2 \times 1}{2 \times 1 \times (5 \times 4 \times 3 \times 2 \times 1)}$	$\dfrac{3!}{2!1!} = \dfrac{3 \times 2 \times 1}{2 \times 1 \times (1)}$
$= \dfrac{7 \times 6}{2 \times 1} = \dfrac{42}{2} = 21$	$= \dfrac{6}{2} = 3$

Now multiply these two values together: $21 \times 3 = 63$ possible ways.

c $\dfrac{63}{210}$

This part uses the information you discovered on the previous two parts. There are 210 ways to select four students for the tennis team. Of those 210, there are 63 ways to select two boys and two girls. Therefore, the probability that the team will be made up of two boys and two girls is $\dfrac{63}{210}$.

d **1**

There's no way that the four-person team could be made up of all girls, because there are only three girls in the group of 10 students. There has to be at least one boy on the team. Whenever something always happens, the probability that it will happen is 1. It's an absolute certainty.

39. *a* 5.2

Before you get started, it's important to note that the diagonals of a rhombus (a) bisect each other, (b) bisect the angles from which they're drawn, and (c) are perpendicular.

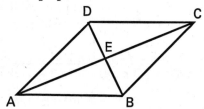

Therefore, △*AED* is a right triangle, the measure of ∠*DAE* is half the measure of ∠*BAD* (or 18°), and *AE* equals half the length of diagonal $\overline{AC}$ (or 8°). To find the length of $\overline{BD}$, first find the length of $\overline{DE}$ using trigonometry, then double it. You know the length of the leg adjacent to ∠*DAE*, and you want to find the length of the opposite leg. Therefore, use tangent (the TOA in SOHCAHTOA):

$$\tan \angle DAE = \frac{DE}{AE}$$

$$\tan 18° = \frac{DE}{8}$$

Since tan 18° = 0.3249, enter that into the equation and cross-multiply:

$$0.3249 = \frac{DE}{8}$$

$$8(0.3249) = DE$$

$$2.5992 = DE$$

The length of diagonal $\overline{BD}$ is twice 2.5992, or 5.1984. When you round this off to the nearest *tenth*, your answer becomes 5.2.

b   34

Finding the perimeter of the rhombus is a little easier. All you have to do is find the length of a side and multiply that number by 4 (since all four sides of a rhombus are equal). Consider $\triangle AED$ once again. You know the length of both legs of this triangle, so find the hypotenuse using the Pythagorean Theorem:

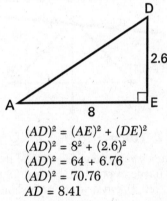

$$(AD)^2 = (AE)^2 + (DE)^2$$
$$(AD)^2 = 8^2 + (2.6)^2$$
$$(AD)^2 = 64 + 6.76$$
$$(AD)^2 = 70.76$$
$$AD = 8.41$$

The perimeter of the rhombus must be 4 times 8.41, or 33.64. When you round your answer off to the nearest integer, it becomes 34.

40. *a*

Piece of cake.

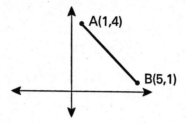

b   **A′(–1,4) and B′(–5,1)**

After a reflection in the *y*-axis, each *y*-coordinate remains the same and each *x*-coordinate is negated. In other words, $r_{y\text{-axis}}(x, y) \rightarrow (-x, y)$. The image of point *A*(1, 4) is *A′*(–1, 4), and the image of point *B*(5, 1) is *B′*(–5, 1). Your graph should look like this:

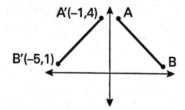

c   **A″(2,8) and B″(10,2)**

When $\overline{AB}$ undergoes a dilation of 2, each of its coordinates is multiplied by 2. The image of point *A*(1,4) is *A″*(2,8), and the image of point *B*(5,1) is *B″*(10,2). Now, your graph should look like this:

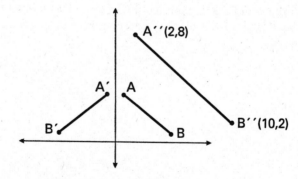

d

What a dumb question. It's obvious that a line segment remains the same length after a reflection. (After all, a reflection of you in a regular mirror is the same size as you are, right?) It's also painfully obvious that if a line segment undergoes a dilation, its size will change. To prove this, find the lengths of $\overline{AB}$, $\overline{A'B'}$, and $\overline{A''B''}$ using the distance formula, which looks like this: $d = \sqrt{(x_2 - x_1)^2 + (y_2 - y_1)^2}$.

$AB = \sqrt{(5 - 1)^2 + (1 - 4)^2}$

$\quad = \sqrt{1^2 + (-3)^2}$

$\quad = \sqrt{16 + 9}$

$\quad = \sqrt{25}$

$\quad = 5$

$A'B' = \sqrt{[-5 - (-1)]^2 + (1 - 4)^2}$

$\quad = \sqrt{(-4)^2 + (-3)^2}$

$\quad = \sqrt{16 + 9}$

$\quad = \sqrt{25}$

$\quad = 5$

So far, you've proved that the reflected segment is the same length as the original one. Now find the length of $\overline{A''B''}$:

$$A''B'' = \sqrt{(10 - 2)^2 + (2 - 8)^2}$$

$$= \sqrt{8^2 + (-6)^2}$$

$$= \sqrt{64 + 36}$$

$$= \sqrt{100}$$

$$= 10$$

Surprise, surprise. After a dilation of 2, the new segment is twice as long as the original one.

Part III

41.

Statements	Reasons
1. ~C (~A ∧ ~B) → C	1. Given
2. ~(~A ∧ ~B)	2. Law of *Modus Tollens*
3. A ∨ B	3. De Morgan's Laws
4. ~A	4. Given
5. B	5. Law of Disjunctive Inference (3, 4)
6. B → D	6. Given
7. D	7. Law of Detachment (5, 6)
8. D → ~E	8. Given
9. ~E	9. Law of Detachment (7, 8)
10. ~F → E	10. Given
11. F	11. Law of *Modus Tollens* (9, 10)

42.

When they don't give you a diagram to refer to, it helps to create one on your own.

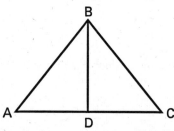

The plan: $\overline{AD}$ and $\overline{CD}$ are corresponding sides of △ABD and △CBD, which share a common side. Prove that the triangles are congruent using SAS, then use CPCTC.

Statements	Reasons
1. $\triangle ABC$ is isosceles with $\overline{AB} \cong \overline{CB}$; $\overline{BD}$ bisects $\angle ABC$	1. Given
2. $\angle ABD \cong \angle CBD$	2. Definition of angle bisector
3. $\overline{BD} \cong \overline{BD}$	3. Reflexive Property of Congruence
4. $\triangle ABD \cong \triangle CBD$	4. SAS $\cong$ SAS
5. $\overline{AD} \cong \overline{CD}$	5. CPCTC

EXAMINATION:
JUNE 1995

Part I

Answer 30 questions from this part. Each correct answer will receive 2 credits. No partial credit will be allowed. Write your answers in the spaces provided on the separate answer sheet. Where applicable, answers may be left in terms of π or in radical form. [60]

1 In the accompanying diagram, $\overleftrightarrow{AB}$ is parallel to $\overleftrightarrow{CD}$, and transversal $\overleftrightarrow{EH}$ intersects $\overleftrightarrow{AB}$ and $\overleftrightarrow{CD}$ at F and G, respectively. If m$\angle AFG = 2x + 10$ and m$\angle FGD = x + 20$, find the value of x.

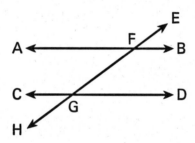

2 In the accompanying diagram, $ABCD$ is a parallelogram, $\overline{DA} \cong \overline{DE}$, and m$\angle B = 70$. Find m$\angle E$.

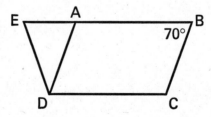

3 In $\triangle ABC$, m $\angle A = 35$ and m $\angle C = 77$. Which is the longest side of the triangle?

4 The sides of a triangle measure 6, 8, and 10. The shortest side of a similar triangle is 15. Find the perimeter of the larger triangle.

5 Rectangle $PROM$ has coordinates $P(2,1)$, $R(8,1)$, $O(8,5)$, and $M(2,5)$. What are the coordinates of the point of intersection of the diagonals?

6 Find, to the *nearest tenth*, the distance between points $(1,3)$ and $(-2,0)$.

7 Solve for x: $\dfrac{2x - 4}{3} = \dfrac{3x + 4}{2}$

8 In the accompanying diagram of right triangle MNQ, $\overline{NP}$ is the altitude to hypotenuse $\overline{MQ}$. If $QP = 16$ and $PM = 9$, find the length of $\overline{NP}$.

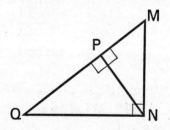

9 How many distinct five-letter permutations can be formed using the letters of the word "GAUSS"?

10 Under a translation, the image of point (3,2) is (−1,3). What are the coordinates of the image of point (−2,6) under the same translation?

11 In $\triangle BAT$, M is the midpoint of $\overline{BA}$ and N is the midpoint of $\overline{BT}$. If $AT = 3x + 12$ and $MN = 15$, find x.

12 How many differeent bowling teams of five persons can be formed from a group of ten persons?

13 A 20-foot ladder is leaning against the wall. The foot of the ladder makes an angle of 58° with the ground. Find, to the *nearest foot*, the vertical distance from the top of the ladder to the ground.

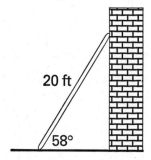

14 In quadrilateral $ABCD$, m $\angle A$ = 57, m $\angle B$ = 65, and m $\angle C$ = 118. What is the measure of an exterior angle at D?

15 Under a dilation with constant of dilation k, the image of the point (2,3) is (8,12). What is the value of k?

Directions (16-34): For *each* question chosen, write the *numeral* preceding the word or expression that best completes the statement or answers the question.

16 An equation of the line that passes through a point $(0,3)$ and whose slope is -2 is

(1) $y = -2x + 3$ (3) $y = 2x + 3$
(2) $y = -2x - 3$ (4) $y = 2x - 3$

17 Given: $p \rightarrow q$

$$\frac{p}{\therefore \quad q}$$

What is the argument called?

(1) DeMorgan's law
(2) Law of attachment
(3) Law of Disjunctive Inference
(4) Law of Contrapositive

18 If $x * y = \dfrac{x^2 - 2xy + y^2}{x - y}$ defines the binary operation *, what is the value of $5 * 3$?

(1) 1   (3) 9
(2) 2   (4) 4

19 If $(x - 3)$ and $(x + 7)$ are the factors of the trinominal $x^2 + ax - 21$, what is the value of a?

(1) -3 (3) 7
(2) -4 (4) 4

20 Which statement is *not* always true about a parallelogram?

    (1) Opposite sides are parallel.
    (2) Opposite sides are congruent.
    (3) Opposite angles are congruent.
    (4) Diagonals are congruent.

21 The parabola shown in the diagram is reflected in the *x*-axis.

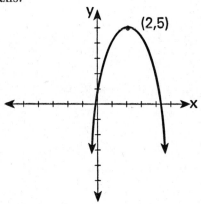

What is the image of the turning point after the reflection?

    (1) (2,–5)   (3) (–2,–5)
    (2) (–2,5)   (4) (2,5)

22 If $\angle C$ is the complement of $\angle A$, and $\angle S$ is the supplement of $\angle A$, which statement is *always* true?

    (1) $m\angle C + m\angle S = 180$
    (2) $m\angle C + m\angle S = 90$
    (3) $m\angle C > m\angle S$
    (4) $m\angle C < m\angle S$

23 Which equation describes the locus of points equidistant from points (2,2) and (2,6)?

(1) $y = 8$ (3) $x = 8$
(2) $y = 4$ (4) $x = 4$

24 In equilateral triangle ABC, the bisectors of angles A and B intersect at point F. What is m$\angle AFB$?

(1) 60   (3) 120
(2) 90   (4) 150

25 Two sides of a triangle have lengths 5 and 8. Which length can *not* be the length of the third side?

(1) 5   (3) 3
(2) 6   (4) 4

26 In the accompanying diagram of right triangle ABC, what is tan C?

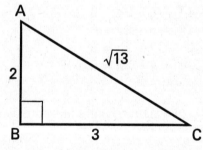

(1) $\dfrac{2}{3}$ (3) $\dfrac{\sqrt{13}}{3}$

(2) $\dfrac{3}{2}$ (4) $\dfrac{2}{\sqrt{13}}$

27 In the accompanying diagram, $\overleftrightarrow{ACE}$ is parallel to $\overleftrightarrow{DB}$, m∠DBA = 40, and m∠BCE = 105.

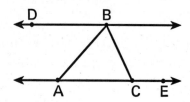

Which statement is true?

(1) $\overline{AB}$ is the longest side of △ABC.

(2) $\overline{AC}$ is the longest side of △ABC.

(3) △ABC is an isosceles triangle.

(4) △ABC is an obtuse triangle.

28 Which equation represents the circle whose center is (–4,2) and whose radius is 3?

(1) $(x + 4)^2 + (y - 2)^2 = 9$

(2) $(x + 4)^2 + (y - 2)^2 = 3$

(3) $(x - 4)^2 + (y + 2)^2 = 9$

(4) $(x - 4)^2 + (y + 2)^2 = 3$

29 If two legs of a right triangle measure 3 and $\sqrt{10}$, then the hypotenuse must measure

(1) 1   (3) 10

(2) $\sqrt{19}$ (4) 19

30 Which statement is equivalent to "If a quadrilateral is a rectangle, the diagonals are congruent"?

(1) If the diagonals of a quadrilateral are congruent, the quadrilateral is a rectangle.
(2) If a quadrilateral is not a rectangle, the diagonals of the quadrilateral are not congruent.
(3) If the diagonals of a quadrilateral are not congruent, the quadrilateral is not a rectangle.
(4) If a quadrilateral is a parallelogram, the diagonals are congruent.

31 In how many points do the graphs of the equations $x^2 + y^2 = 9$ and $y = 2x - 1$ intersect?

(1) 1 (3) 3
(2) 2 (4) 4

32 Which quadratic equation has irrational roots?

(1) $x^2 + 2x - 8 = 0$ (3) $x^2 - 3x + 2 = 0$
(2) $x^2 - x - 30 = 0$ (4) $x^2 - 4x - 7 = 0$

33 Which equation represents the axis of symmetry of the graph of the equation $y = x^2 - 6x + 5$?

(1) $x = -3$ (3) $x = 3$
(2) $y = -3$ (4) $y = 3$

34 Which equation represents a line that is parallel to the line whose equation is $y = \frac{1}{2}x - 2$?

(1) $y = 2x - 3$ (3) $2y = x - 3$
(2) $y = -2x - 3$ (4) $2y = -x - 3$

Directions (35): Show all costruction lines.

35 Construct the angle bisector of $\angle C$ of $\triangle ABC$

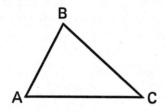

Part II

Answer *three* questions from this part. Clearly indicate the necessary steps, including appropriate formula substitutions, diagrams, graphs, charts, etc. Calculations that may be obtained by mental arithmetic or the calculator do not need to be shown. [30]

36 *a* On graph paper, draw the graph of the equation $y = x^2 - 2x - 3$ for all values of x in the interval $-2 \leq x \leq 4$.

b What are the roots of the equation $x^2 - 2x - 3 = 0$?

c On the same set of axes, draw the image of the graph drawn in part *a* after a reflection in the y-axis.

37 Answer both *a* and *b* for all values of *x* for which these expressions are defined.

a Simplify: $\dfrac{x^2 + 9x + 20}{x^2 - 16} \div \dfrac{x^2 + 5x}{4x - 16}$

b Solve for *x*: $\dfrac{2}{x} = \dfrac{x - 3}{5}$

38 A debating team of four persons is to be chosen from five juniors and three seniors.

a How many different four-member teams are possible? [2]

b How many of these teams will consist of exactly two juniors and two seniors? [3]

c What is the probability that one of the four-member teams will consist of exactly one junior and three seniors? [3]

d What is the probability that one four-member team will consist of juniors only? [2]

39 In the accompanying diagram of right triangle *ABD*, *AB* = 6 and altitude $\overline{BC}$ divides hypotenuse $\overline{AD}$ into segments of lengths *x* and 8.

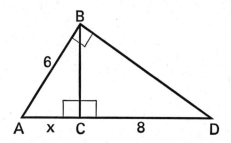

   a Find *AC* to the *nearest tenth*.   [7]

   b Using the answer from part *a*, find the measure of ∠*A* to the *nearest degree*.   [3]

40 An 8- by 10-inch photo has a frame of uniform width placed around it.

   a If the uniform width of the frame is *x* inches, express the outside dimensions of the picture frame in terms of *x*.   [4]

   b If the area of the picture and frame is 143 in²· what is the uniform width of the frame?   [6]

Part III

Answer *one* question from this part. Clearly indicate the necessary steps, including appropriate formula substitutions, diagrams, graphs, charts, etc. Calculations that may be obtained by mental arithmetic or the calculator do not need to be shown. [10]

41 *a* Given: Either I go to camp or I get a summer job.

If I get a summer job, then I will earn money.

If I earn money, then I will buy new sneakers.

I do not buy new sneakers.

Let *C* represent: "I go to camp."

Let *J* represent: "I get a summer job."

Let *M* represent: "I earn money."

Let *S* represent: "I buy new sneakers."

Prove: I go to camp. [8]

b Given the true statements:

If Michael is an athlete and he is salaried, then Michael is a professional.

Michael is not a professional.

Michael is an athlete.

Which statement must be true? [2]

(1) Michael is an athlete and he is salaried.
(2) Michael is a professional or he is salaried.
(3) Michael is not salaried.
(4) Michael is not an athlete.

42 Given: quadrilateral $PQRT$, $\overline{QSV}$, $\overline{RST}$, $\overline{PTV}$,

$\overline{QV}$ bisects $\overline{RT}$, and $\overline{QR} \parallel \overline{PV}$.

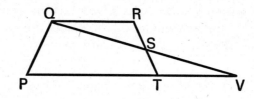

Prove: $\overline{QS} \cong \overline{VS}$ [10]

ANSWER KEY

Part I

1. 10	13. 17	25. (3)
2. 70	14. 60°	26. (1)
3. $\overline{AB}$	15. 4	27. (1)
4. 60	16. (1)	28. (1)
5. (5, 3)	17. (2)	29. (2)
6. 4,2	18. (2)	30. (3)
7. −4	19. (4)	31. (2)
8. 12	20. (4)	32. (4)
9. 60	21. (1)	33. (3)
10. (−6, 7)	22. (4)	34. (3)
11. 6	23. (2)	35. construction
12. 252	24. (3)	

EXPLANATIONS:
JUNE 1995

Part I

1. 10

Since $\overline{AB}$ is parallel to $\overline{CD}$, $\angle AFG$ and $\angle FGD$ are alternate interior angles (which have the same measure). Set them equal to each other and solve for x:

$$2x + 10 = x + 20$$
$$x = 10$$

2. 70

Since $ABCD$ is a parallelogram, you know that AD is parallel to BC. That means $\angle EAD$ and $\angle ABC$ are corresponding angles. Therefore, they both measure 70°. Now turn your attention to $\triangle ADE$:

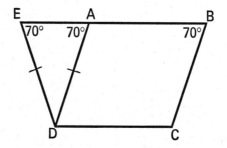

Since $\overline{DA} \cong \overline{DE}$, $\triangle ADE$ is isosceles. If two sides of a triangle are congruent, the angles opposite those two sides are also congruent. Thus, m$\angle E$ = m$\angle EAD$ = 70

3. $\overline{AB}$

Your diagram should look something like this:

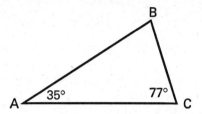

From the Rule of 180, the sum of the three angles in a triangle is always 180°. Therefore:

$$m \angle A + m \angle B + m \angle C = 180$$
$$35 + m \angle B + 77 = 180$$
$$m \angle B = 68$$

In any triangle, the largest side is opposite the largest angle. Since $\angle C$ is the biggest angle, the side opposite it, $\overline{AB}$, has the greatest length.

4. 60

Any time a problem involves similar triangles, all you need to do is set up a proportion. The key is lining up the corresponding sides:

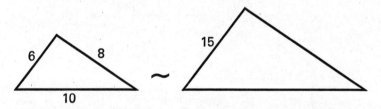

From the diagram, you can see that the shortest side of the smaller triangle is 6 units long, and its counterpart of the larger triangle is 15 units long. The perimeter of the smaller triangle is 6 + 8 + 10, or 24, and you're looking for the perimeter of the larger triangle (call it P). Your proportion should look like this:

$$\frac{6}{15} = \frac{24}{P}$$

Cross-multiply and solve:

$$6P = 360$$
$$P = 60$$

5. (5,3)

This problem might look pretty tough unless you realize that the diagonals of a rectangle bisect each other. Therefore, they each intersect at each other's midpoint. All you have to do is find the midpoint of one of the diagonals, and you'll have the point of intersection. Use the midpoint formula:

$$(\bar{x}, \bar{y}) = \left(\frac{x_1 + x_2}{2}, \frac{y_1 + y_2}{2} \right)$$

Making a diagram helps you decide which points to use in the midpoint formula. (Be sure not to select two adjacent points by accident!)

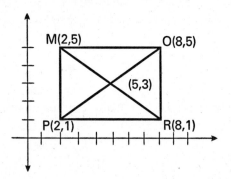

Choose $R(8,1)$ and $M(2,5)$:

$$\bar{x} = \frac{8+2}{2} = \frac{10}{2} = 5$$

$$\bar{y} = \frac{1+5}{2} = \frac{6}{2} = 3$$

The midpoint of diagonal $\overline{RM}$ is (5,3).

6. 4.2

Use the Distance Formula to find the distance between two points:

$$d = \sqrt{(y_2 - y_1)^2 + (x_2 - x_1)^2}$$

Let (x_1, y_1) equal $(1,3)$, and (x_2, y_2) equal $(-2,0)$:

$$d = \sqrt{(0 - 3)^2 + (-2 - 1)^2} = \sqrt{(-3)^2 + (-3)^2} = \sqrt{9 + 9} = \sqrt{18} = 4.2$$

Be sure your answer is expressed to the nearest *tenth*, as the question specifies.

7. –4

Any time two fractions are equal to each other, you can cross-multiply:

$$\frac{2x - 4}{3} = \frac{3x + 4}{2}$$
$$2(2x - 4) = 3(3x + 4)$$
$$4x - 8 = 9x + 12$$
$$-5x = 20$$
$$x = -4$$

To check your math, plug –4 in for x in the original equation and make sure it works.

8. 12

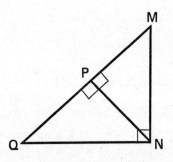

This figure represents three similar right triangles (ΔMPN, ΔNPQ, and ΔMNQ), and all of their corresponding sides are proportional to each other. For this problem, consider ΔMPN and ΔNPQ and set up a proportion. In ΔMNP, NP is the long leg and PM is the short leg; in ΔNPQ, QP is the long leg and NP is the short leg. Therefore:

$$\frac{PM}{NP} = \frac{NP}{QP}$$

$$\frac{9}{NP} = \frac{NP}{16}$$

When you cross-multiply, you'll get:

$$(NP)^2 = 144$$
$$NP = \{12, -12\}$$

Since you're looking for a distance, which must be positive, forget about −12.

9. 60

If GAUSS didn't have any duplicate letters, the answer to this question would be simple; a five-letter word has 5! possible permutations. The only slight difference involves accounting for the repetition. To find the number of possible arrangements of the letters in a word with n letters, in which one letter appears p times (remember that p is greater than 1), the formula looks like this:

$$\frac{n!}{p!}$$

GAUSS has five letters, but there are two S's. Therefore, you can express the number of arrangements as:

$$\frac{5!}{2!} = \frac{5 \times 4 \times 3 \times 2 \times 1}{2 \times 1} = 5 \times 4 \times 3 = 60$$

10. (−6, 7)

The translation that maps the point (3, 2) onto (−1, 3) subtracts 4 from the x-coordinate (because $3 - 4 = -1$) and adds 1 to the y-coordinate (because $2 + 1 = 3$). You can write the translation like this:

$$(x, y) \rightarrow (x - 4, y + 1)$$

Under this same translation, the point (−2, 6) is mapped onto (−2 − 4, 6 + 1), or (−6, 7).

11. 6

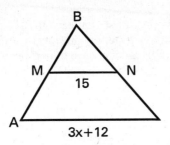

A segment that connects the midpoints of two sides of a triangle is half the length of the third side. Since M is the midpoint of $\overline{BA}$ and N is the midpoint of $\overline{BT}$, MN must be half of AT. Set up the equation:

$$MN = \frac{1}{2}(AT)$$

$$15 = \frac{1}{2}(3x + 12)$$

Multiply both sides by 2 to get rid of the fraction, then solve:

$$30 = 3x + 12$$
$$18 = 3x$$
$$6 = x$$

12. 252

This problem involves combinations, because the order in which the bowlers are chosen doesn't matter. Therefore, use the Combinations Formula:

$$_nC_r = \frac{n!}{r!(n-r)!}$$

There are 10 people in the group ($n = 10$), and you want to choose 5 of them ($r = 5$):

$$_{10}C_5 = \frac{10!}{5!(10-5)!} = \frac{10 \times 9 \times 8 \times 7 \times 6 \times 5 \times 4 \times 3 \times 2 \times 1}{5 \times 4 \times 3 \times 2 \times 1 \times (5 \times 4 \times 3 \times 2 \times 1)} = 252$$

13. 17

It's trigonometry time. First, you have to decide which of the three trig functions (sine, cosine, or tangent) to use. You've been given the hypotenuse (20), and you have to find the length of the side opposite the angle, so use sine (the SOH in SOHCAHTOA):

$$\sin 58° = \frac{x}{20}$$

$$0.8480 = \frac{x}{20}$$

$$20(0.8480) = x$$

$$16.96 = x$$

Now, round the result to the nearest foot, as the question specifies. Your answer is 17.

14. 60°

Draw *ABCD* first. It should look something like this:

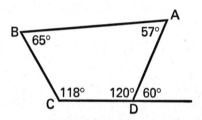

In order to figure out the measure of the exterior angle ($\angle CDE$), you have to figure out the measure of interior angle $\angle ADC$. The sum of the four angles in a quadrilateral must always be 360°, so:

$$m\angle A + m\angle B + m\angle C + m\angle ADC = 360$$
$$57 + 65 + 118 + m\angle ADC = 360$$
$$m\angle ADC = 120$$

Since $\angle ADC$ and $\angle CDE$ are supplementary, their sum must be 180°. Thus:

$$m\angle ADC + m\angle CDE = 180$$
$$120 + m\angle CDE = 180$$
$$m\angle CDE = 60$$

15. 4

When a figure undergoes a dilation, each point is multiplied by a certain value and the resulting figure is either smaller or larger than the original. In this question, each of the coordinates in the point (2,3) has been multiplied by 4; the image is (8,12). Therefore, the value of k is 4.

Multiple Choice

16. (1)

Here's a standard test of your knowledge of the $y = mx + b$ format. The m represents the slope of the line and the b represents the y-intercept. Eliminate answer choices (3) and (4) because the slope of those two lines is 2. Further, (0,3) is the y-intercept of the line because the x-coordinate is 0; b must equal 3. The correct equation is $y = -2x + 3$.

17. (2)

If you remember that the contrapositive is the same thing as the Flip- and Negate-Rule, you can eliminate answer choice (4). Further, the example is much too simple to involve De Morgan's Laws, which involve the negation of conjunctions ($p \wedge q$) and disjunctions ($p \vee q$). Eliminate answer choice (1). The Law of Detachment simply states that if $p \rightarrow q$ and p is true, then q must also be true.

18. (2)

Here's another function example in which there's a wacky symbol designed to confuse you. The function "$\circ$" is defined, so plug in $x = 5$ and $y = 3$:

$$x \circ y = \frac{x^2 - 2xy + y^2}{x - y}$$

$$5 \circ 3 = \frac{5^2 - 2(5)(3) + 3^2}{5 - 3} = \frac{25 - 30 + 9}{2} = \frac{4}{2} = 2$$

19. (4)

The best way to figure this out is to combine the two factors you've been given using FOIL:

$$(x - 3)(x + 7) = x^2 + 7x - 3x - 21 = x^2 + 4x - 21$$

From the expanded expression, it must be true that $a = 4$.

20. (4)

Each of the first three answer choices is true for all parallelograms, but answer choice (4) is not. For example, you might see a parallelogram that looks like this:

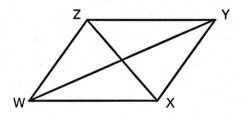

Notice that diagonal $\overline{WY}$ is much longer than diagonal $\overline{XZ}$. The only time the diagonals of a parallelogram are congruent is when that parallelogram is a rectangle or a square.

21. (1)

After a reflection in the x-axis, the x-coordinate remains the same and the y-coordinate is negated. In other words, $r_{x\text{-axis}}(x, y) \to (x, -y)$. Therefore, the image of the point $(2,5)$ is $(2,-5)$.

Note: Another way to solve this problem is to plot all four points on the given diagram, like so:

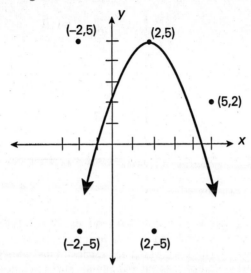

Note that answer choices (2), (3), and (4) aren't even close.

22. **(4)**

This problem involves a lot of variables. (How big is each angle?) If you plug in values for the three angles, you'll make the problem a lot easier to deal with. Let m $\angle A$ = 30. Since $\angle C$ is complementary to $\angle A$, then m $\angle C$ = 60. Further, since the sum of any angle and its supplement is 180°, m $\angle S$ = 150. Try these numbers in all four answer choices; you'll find that only answer choice (4) works.

23. (2)

Plot the two points on the coordinate axes, like so:

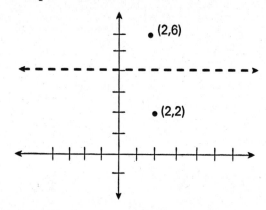

From the graph, you can see that the locus of points that are equidistant from these two points is a horizontal line, so cross off answer choices (3) and (4), which are vertical. Plus, you know that the line has to run in between the two points; answer choice (1) can't be right, because it runs above both points.

The rule to remember: the locus of all points equidistant from two points is the perpendicular bisector of the segment between those two points.

24. (3)

Draw equilateral △*ABC* and the angle bisectors $\overline{AD}$ and $\overline{BE}$, which intersect at point *F*:

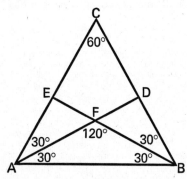

Look at ∠AFB. Is it acute or obtuse? It looks much bigger than 90°, so it's obtuse. You can cross off answer choices (1) and (2) because they're too small.

Now consider ∆AFB. Both ∠CAB and ∠CBA measure 60° because the triangle is equilateral. Since $\overline{AD}$ and $\overline{BE}$ are angle bisectors, ∠BAF and ∠ABF measure 30° each. The sum of the three angles in a triangle is 180°, so m∠AFB = 120.

25. (3)

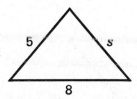

When you're given the lengths of two sides of a triangle, you know that the third side has to be shorter than the sum and larger than the difference of the other two sides. In this case, the sides of the triangle measure 5, 8, and s. Therefore:

$$8 - 5 < s < 8 + 5$$
$$3 < s < 13$$

The only answer choice that doesn't fall within this range is answer choice (3).

26. (1)

The tangent function involves the TOA in SOHCAHTOA. Thus, the tangent of an angle equals $\dfrac{opposite}{adjacent}$:

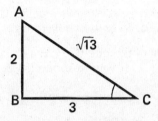

The length of the side opposite $\angle C$ is 2, and the length of the adjacent side is 3. The tangent must therefore be $\frac{2}{3}$.

27. (1)

To answer this question, you have to figure out the measure of each angle in $\triangle ABC$. Finding the measure of $\angle BCA$ is the easiest; since m $\angle BCE = 105$, and $\angle BCA$ and $\angle BCE$ are supplementary, m $\angle BCA = 75$. To find the measure of $\angle BAC$, think only of the two parallel lines and the transversal $\overline{AB}$, like so:

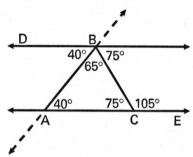

$\angle DBA$ and $\angle BAC$ are alternate interior angles, so they have the same measure. Therefore, m $\angle BAC = 40$. Since the sum of the measures of all three angles in a triangle is 180°, m $\angle ABC = 65$.

Since $\angle BCA$ is the biggest angle in the triangle, the side opposite that angle, $\overline{AB}$, must be the biggest side of $\triangle ABC$.

28. (1)

The formula for a circle with center (h, k) and radius r is:

$$(x - h)^2 + (y - k)^2 = r^2$$

Since the radius of the circle in this problem is 3, you know that the equation has to equal 3^2, or 9. Eliminate answer choices (2) and (4).

The key to the rest of this question is remembering the minus signs in the formula above. If you substitute $(-4,2)$ for (h, k) in the formula, you get:

$$[x - (-4)]^2 + (y - 2)^2 = 9$$
$$(x + 4)^2 + (y - 2)^2 = 9$$

29. (2)

First, draw your right triangle. Since the question specifies that the two *legs* measure 3 and $\sqrt{10}$, you can place the measurements like so:

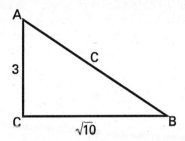

Now it's just a matter of using the Pythagorean Theorem:

$$a^2 + b^2 = c^2$$
$$(3)^2 + (\sqrt{10})^2 = c^2$$
$$9 + 10 = c^2$$
$$19 = c^2$$
$$\sqrt{19} = c$$

30. (3)

The statement in this problem is a simple conditional statement. Let R represent "a quadrilateral is a rectangle" and C represent "the diagonals are congruent." You can now diagram the statement like this: $R \rightarrow C$.

First, eliminate answer choice (4), because the statement in the question doesn't mention parallelograms. Now diagram the other three answer choices:

(1) $C \rightarrow R$

(2) $\sim R \rightarrow \sim C$

(3) $\sim C \rightarrow \sim R$

If you flip and negate the original statement, you'll get answer choice (3)—the contrapositive inference

31. (2)

Since you don't have to identify the exact coordinates of the points of intersection, don't bother launching into any hard-core algebra. The easiest way to determine the number of intersection points is to draw both graphs. The first equation, $x^2 + y^2 = 9$, is a circle with a radius of 3 centered at the origin. The graph of $y = 2x - 1$ is a line that intersects the y-axis at $(0,-1)$ and has a slope of 2. Your graph should look like this:

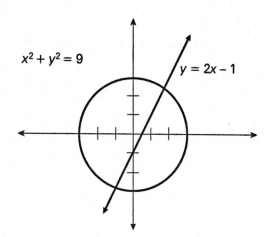

There are two points of intersection.

32. (4)

If you can factor a quadratic equation, you know that its roots are rational. Try to factor each of the four answer choices; the one you can't factor has irrational roots.

(1) $x^2 + 2x - 8 = 0$
     $(x + 4)(x - 2) = 0$
     $x = \{-4, 2\}$

(2) $x^2 - x - 30 = 0$
     $(x - 6)(x + 5) = 0$
     $x = \{6, -5\}$

(3) $x^2 - 3x + 2 = 0$
     $(x - 2)(x - 1) = 0$
     $x = \{2, 1\}$

(4) Not factorable.

You have to use the Quadratic Formula to get the roots of answer choice (4). (By the way, those roots are $2 + \sqrt{11}$ and $2 - \sqrt{11}$, in case you're curious.)

33. (3)

Since the parabola in this question is in standard form $(y = ax^2 + bx + c)$, you can find the axis of symmetry using the equation:

$$x = -\frac{b}{2a}$$

The equation is $y = x^2 - 6x + 5$, so $a = 1$, $b = -6$, and $c = 5$. Therefore, the equation for the axis of symmetry is:

$$x = -\frac{-6}{2(1)} = \frac{6}{2} = 3$$

The line $x = 3$ is the axis of symmetry.

34. (3)

All parallel lines have the same slope. The line in the question is in $y = mx + b$ format, so you know its slope is $\frac{1}{2}$. Therefore, you can eliminate answer choices (1) and (2) because neither has a slope of $\frac{1}{2}$.

Answer choices (3) and (4) are not in $y = mx + b$ format; since each begins with $2y$, you have to divide both equations by 2 in order for y to be alone on the left side of the equal sign. Thus, in answer choice (3):

$$2y = x - 3$$
$$y = \frac{1}{2}x - \frac{3}{2}$$

Now that the equation is in proper format, you can see that the slope of this line is also $\frac{1}{2}$.

35.

Put the metal point of the compass on point C and create an arc that intersects both $\overline{AC}$ and $\overline{BC}$, and call those points of intersection D and F:

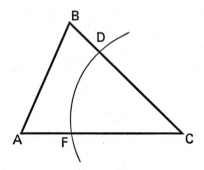

Without changing the width of your compass, put the pointy end on point D and make an arc inside the angle. Then put the pointy end on point F and repeat Step 2, and find the point where the two arcs intersect (label this point E):

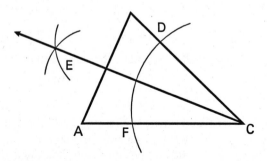

Draw ray CE. This is the angle bisector.

Part II

36. *a*

Start the graphing process by plugging the integers between –2 and 4, inclusive, into the equation for the parabola and determine its coordinates. For example, if $x = -2$, then $y = (-2)^2 - 2(-2) - 3$, or 5. Your first point is (–2, 5). Here's the rest of the T-chart, as well as the graph of the parabola:

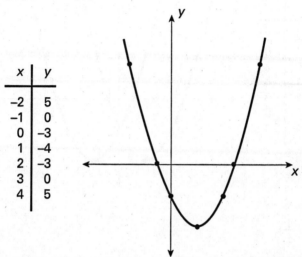

x	y
–2	5
–1	0
0	–3
1	–4
2	–3
3	0
4	5

b $x = \{3, -1\}$

The roots of a parabolic equation tell you where it intersects the x-axis (where $y = 0$). To find the roots of this equation, factor it and solve for x:

$$x^2 - 2x - 3 = 0$$
$$(x - 3)(x + 1) = 0$$
$$x = \{3, -1\}$$

Check your work by plugging –3 and 1 back into the equation and making sure they work. (Or, check your graph to see where it intersects the x-axis.)

c

After a reflection in the y-axis, each y-coordinate remains the same and each x-coordinate is negated. In other words, $r_{y\text{-axis}}(x, y) \rightarrow (-x, y)$. Convert the points in the parabola as shown below; your graph should look like this:

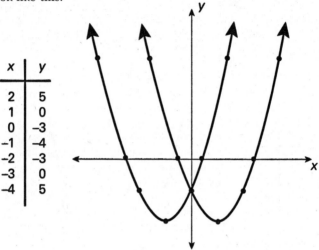

x	y
2	5
1	0
0	-3
-1	-4
-2	-3
-3	0
-4	5

37. a $\dfrac{4}{x}$

The problem is presented as a division problem, so turn it into a multiplication problem by flipping the second term:

$$\frac{x^2 + 9x + 20}{x^2 - 16} \cdot \frac{4x - 16}{x^2 + 5x}$$

Now factor all the complex terms like this:

$$x^2 + 9x + 20 = (x + 4)(x + 5)$$
$$x^2 - 16 = (x + 4)(x - 4)$$
$$4x + 16 = 4(x + 4)$$
$$x^2 + 5x = x(x + 5)$$

Once you've factored these four terms, the problem looks like this:

$$\frac{(x + 4)(x + 5)}{(x + 4)(x - 4)} \cdot \frac{4(x - 4)}{x(x + 5)}$$

Cancel out all the factors that appear both on the top and on the bottom, and you're left with:

$$\frac{\cancel{(x+4)}\cancel{(x+5)}}{\cancel{(x+4)}\cancel{(x-4)}} \cdot \frac{4\cancel{(x-4)}}{x\cancel{(x+5)}} = \frac{4}{x}$$

b $x = \{5, -2\}$

Whenever two fractions are equal to each other, you can cross-multiply:

$$\frac{2}{x} = \frac{x-3}{5}$$
$$x(x-3) = 2 \times 5$$
$$x^2 - 3x = 10$$
$$x^2 - 3x - 10 = 0$$

Now factor the polynomial and set each factor equal to zero:

$$(x-5)(x+2) = 0$$
$$x = \{5, -2\}$$

Plug these two values back into the equation to make sure they work.

38. a 70

This problem involves combinations, because the order of the debaters doesn't matter. If you're finding the number of combinations of r people and you can choose from a group of n people, the formula is:

$$_nC_r = \frac{n!}{r!\,(n-r)!}$$

To find the number of ways you can choose four debating team members from a group of eight (five juniors and three seniors), you need to find the value of $_8C_4$ which equals:

$$\frac{8!}{4!\,6!} = \frac{8 \times 7 \times 6 \times 5 \times 4 \times 3 \times 2 \times 1}{4 \times 3 \times 2 \times 1 \times (4 \times 3 \times 2 \times 1)}$$
$$= \frac{8 \times 7 \times 6 \times 5}{4 \times 3 \times 2 \times 1} = \frac{1,680}{24} = 70$$

_b_ 30

This one's a bit more complicated. You want to find the number of ways 2 out of 5 juniors can be chosen ($_5C_2$) and the number of ways 2 out of 3 seniors can be chosen ($_3C_2$):

$$\frac{5!}{2!\,3!} = \frac{5 \times 4 \times 3 \times 2 \times 1}{2 \times 1 \times (3 \times 2 \times 1)} \qquad\qquad \frac{3!}{2!\,1!} = \frac{3 \times 2 \times 1}{2 \times 1 \times (1)}$$

$$= \frac{5 \times 4}{2 \times 1} = \frac{20}{2} = 10 \qquad\qquad\qquad = \frac{6}{2} = 3$$

Now multiply these two values together: $10 \times 3 = 30$ possible ways.

_c_ $\dfrac{5}{70}$

Now you need to find out how many of those teams will consist of 1 junior and 3 seniors. You can do this by finding the number of ways 1 out of 5 juniors can be chosen ($_5C_1$) and the number of ways 3 out of 3 seniors can be chosen ($_3C_3$):

$$\frac{5!}{1!\,4!} = \frac{5 \times 4 \times 3 \times 2 \times 1}{1 \times (4 \times 3 \times 2 \times 1)} \qquad\qquad \frac{3!}{3!\,0!} = \frac{3 \times 2 \times 1}{3 \times 2 \times 1 \times (1)}$$

$$= \frac{5}{1} = 5 \qquad\qquad\qquad\qquad = 1$$

Now multiply these two values together: $5 \times 1 = 5$ possible teams with 1 junior and 3 seniors. From Part A, you know there are 70 possible 4-student teams. Therefore, the chance that a team will have

1 junior and 3 seniors is $\dfrac{5}{70}$.

_d_ $\dfrac{5}{70}$

To find number of 4-person teams made up of 4 juniors, find the number of ways 4 juniors can be chosen from a group of 5 ($_5C_4$):

$$\frac{5!}{4!\,1!} = \frac{5 \times 4 \times 3 \times 2 \times 1}{(4 \times 3 \times 2 \times 1) \times 1} = \frac{5}{1} = 5$$

There are 5 possible teams made up only of juniors. From Part A, you know there are 70 possible 4-student teams. Therefore, the chance that a team will be made up only of juniors is also $\frac{5}{70}$.

39. *a*   3.2

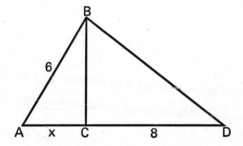

This figure represents three similar right triangles ($\triangle ABC$, $\triangle BDC$, and $\triangle BAD$), and all of their corresponding sides are proportional to each other. For this problem, consider $\triangle ABC$ and $\triangle BAD$ and set up a proportion. In $\triangle ABC$, AC is the short leg and AB is the hypotenuse; in $\triangle BAD$, AB is the short leg and AD is the hypotenuse. Therefore:

$$\frac{AC}{AB} = \frac{AB}{AD}$$

$$\frac{x}{6} = \frac{6}{x+8}$$

When you cross-multiply, you'll get:

$$x(x+8) = 6 \times 6$$

$$x^2 + 8x = 36$$

$$x^2 + 8x - 36 = 0$$

Unfortunately, you can't factor this. (The fact that they want the answer to the nearest *tenth* is a pretty big hint.) You have to use the Quadratic Formula:

$$x = \frac{-b \pm \sqrt{b^2 - 4ac}}{2a}$$

In the equation $x^2 + 8x - 36 = 0$, $a = 1$, $b = 8$, and $c = -36$:

$$x = \frac{-8 \pm \sqrt{(8)^2 - 4(1)(-36)}}{2(1)}$$

$$= \frac{-8 \pm \sqrt{64 + 144}}{2}$$

$$= \frac{-8 + \sqrt{208}}{2}, \frac{-8 - \sqrt{208}}{2}$$

Use your calculator to determine that $\sqrt{208} = 14.42$, and substitute it into the roots:

$$x = \frac{-8 + 14.42}{2} \qquad\qquad x = \frac{-8 - 14.42}{2}$$

$$= \frac{6.42}{2} \qquad\qquad\qquad = \frac{-22.42}{2}$$

$$= 3.21 \qquad\qquad\qquad\quad = -11.21$$

You're looking for a distance, which can't be negative; throw out −11.21. When you round the positive root off to the nearest tenth, your answer becomes 3.2.

b 58

Now it's time for the trigonometry. Consider $\triangle ABC$; you want to find the measure of $\angle A$, and you know the length of the hypotenuse (6) and the adjacent leg (3.2, from Part A). Therefore, you should use the cosine (the CAH in SOHCAHTOA):

$$\cos \angle A = \frac{AC}{AB}$$

$$\cos \angle A = \frac{3.2}{6}$$

$$\cos \angle A = 0.5333$$

Enter 0.5333 into your calculator and press the inverse cosine button (it usually says "cos^{-1}" and involves the "second function" key). You get 57.77. (If you didn't get this, make sure your calculator is in "degree" mode.) When you round this off to the nearest degree, as instructed, you get 58°.

40. *a* $w = 8 + 2x, l = 10 + 2x$

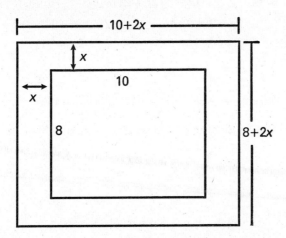

The drawing says it all. The dimensions of the picture are 8×10, and there is a frame around it of uniform width, x. The width of the entire frame is $8 + 2x$, because the frame is on *both sides* of the picture. (Some students get careless and forget that.) Similarly, the length of the whole thing is $10 + 2x$.

b $\dfrac{3}{2}$ or 1.5

The formula for the area of a rectangle is $l \times w$. You know the area, so you have to solve for x:

$$(8 + 2x)(10 + 2x) = 143$$

Using FOIL, you get:

$$80 + 16x + 20x + 4x^2 = 143$$

Combine the terms and set the quadratic equal to zero:

$$4x^2 + 36x + 80 = 143$$
$$4x^2 + 36x - 63 = 0$$

It's nasty, but it's also factorable:

$$(2x - 3)(2x + 21) = 0$$

$2x - 3 = 0$	$2x + 11 = 0$
$2x = 3$	$2x = -11$
$x = \dfrac{3}{2}$	$x = -\dfrac{11}{2}$

The width of the frame has to have a positive value, so get rid of $-\dfrac{11}{2}$. The width of the frame must be $\dfrac{3}{2}$, or 1.5, inches.

Part III

41. *a*

Step 1: Turn all the givens into symbolic terms:

"Either I go to camp or I get a summer job." $(C \vee J)$

"If I get a summer job, then I will earn money." $J \rightarrow M$

"If I earn money, then I will buy new sneakers." $M \rightarrow S$

"I buy new sneakers." S

Step 2: Decide what you want to prove:

"I do not buy new sneakers." $\sim S$

Step 3: Write the proof.

Statements	Reasons
1. $M \rightarrow S$; $\sim S$	1. Given
2. $\sim M$	2. Law of *Modus Tollens*
3. $J \rightarrow M$	3. Given
4. $\sim J$	4. Law of *Modus Tollens* (2, 3)
5. $(C \vee J)$	5. Given
6. C	6. Law of Disjunctive Inference (5,6)

b **(3)**

Symbolize the terms like this:

Let A = "Michael is an athlete."

Let P = "Michael is professional."

Let S = "Michael is salaried."

Now for your symbolization of the statements:

"If Michael is an athlete and he is salaried, then Michael is a professional." $(A \wedge S) \rightarrow P$

"Michael is not a professional." $\sim P$

"Michael is an athlete." A

Use the Law of *Modus Tollens* on the first two statements: if $(A \wedge S) \rightarrow P$ and $\sim P$, then $\sim(A \wedge S)$.

Using De Morgan's Laws, $\sim(A \wedge S)$ converts to $\sim A \vee \sim S$. In English, this means "either Michael is not an athlete OR he is not salaried." You know from the third statement that Michael is an athlete (A), so it must be true (from the Law of Disjunctive Inference) that Michael is not salaried.

42.

The plan: $\overline{QS}$ and $\overline{VS}$ are corresponding sides of $\triangle QSR$ and $\triangle TSV$, which have a vertical angle in common. Prove the triangles congruent using ASA, then use CPCTC.

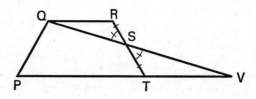

Prove: $\overline{QS} \cong \overline{VS}$

Statements	Reasons
1. $\overline{QR}$ is parallel to $\overline{PT}$	1. Given
2. $\angle R \cong \angle STV$	2. Alternate interior angles are congruent
3. $\overline{QV}$ bisects $\overline{RT}$	3. Given
4. $\overline{RS} \cong \overline{ST}$	4. Definition of bisect
5. $\angle RSV \cong \angle TSV$	5. Vertical angles are congruent
6. $\triangle QRS \cong \triangle VTS$	6. ASA $\cong$ ASA
7. $\overline{QS} \cong \overline{VS}$	7. CPCTC

EXAMINATION: JANUARY 1996

Part I

Answer 30 questions from this part. Each correct answer will receive 2 credits. No partial credit will be allowed. Write your answers in the spaces provided on the separate answer sheet. Where applicable, answers may be left in terms of π or in radical form. [60]

1 If $a \, \Delta \, b$ is a binary operation defined as $\dfrac{2a + b}{a}$, evaluate $2 \, \Delta \, 4$.

2 In the accompanying diagram, $\triangle ABC$ is isosceles, $\overline{BC}$ is extended to D, $\overline{AB} \cong \overline{AC}$, and m $\angle A$ = 80. Find m $\angle ACD$.

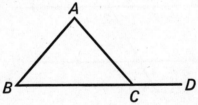

3 In $\triangle PEN$, m $\angle P$ = 40 and $\angle N$ = 80. Which side of the triangle is the longest?

4 Solve for x: $\dfrac{x - 2}{x + 4} = \dfrac{x + 2}{x + 12}$

5 The length of the sides of a triangle are 4, 5, and 6. If the length of the longest side of a similar triangle is 15, what is the length of the *shortest* side of this triangle?

6 The coordinates of the vertices of △ABC are A(0,0), B(3,0), and C(0,4). What is the length of $\overline{BC}$?

7 What is the slope of the line determined by points (–1,3) and (3,–1)?

8 What are the coordinates of N′, the image of N(5,–3) under a reflection in the y-axis?

9 In the accompanying diagram, line *l* is parallel to line *m* and line *t* is a transversal. If m ∠1 = 2x + 20 and m ∠2 = 4x + 10, what is the number of degrees in ∠3?

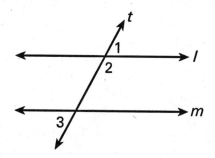

10 Find the number of square units in the area of the triangle whose vertices are points A(2,0), B(6,0), and C(8,5).

11 The measure of one angle of a triangle equals the sum of the measures of the other two angles. Find the number of degrees in the measure of the largest angle of the triangle.

12 In the accompanying diagram, $\overleftrightarrow{AB} \parallel \overleftrightarrow{CD}$, m∠x = 50, and m∠y = 60. What is m∠z?

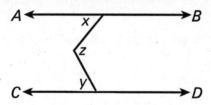

13 A translation maps A(–3,4) onto A′(2,–6). Find the coordinates of B′, the image of B(–4,0) under the same translation.

14 How many committees of three students can be chosen from a class of seven students?

15 In the accompanying diagram of right triangle ABC, b = 40 centimeters, m∠A = 60, and m∠C = 90. Find the number of centimeters in the length of side c.

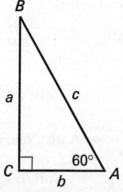

16 Factor completely: $3x^2 - 15x - 42$

Directions (17-35): For *each* question chosen, write the *numeral* preceding the word or expression that best completes the statement or answers the question.

17 Which statement is the negation of $p \wedge \sim q$?

(1) $\sim p \vee q$ (3) $p \wedge q$

(2) $p \vee \sim q$ (4) $\sim p \wedge \sim q$

18 Which equation illustrates the additive inverse property?

(1) $a + (-a) = 0$ (3) $a \div (-a) = -1$

(2) $a + 0 = a$ (4) $a \bullet \dfrac{1}{a} = 1$

19 In right triangle ABC, altitude $\overline{CD}$ is drawn to hypotenuse $\overline{AB}$. If $AD = 2$ and $DB = 6$, then AC is

(1) $4\sqrt{3}$ (3) 3

(2) $2\sqrt{3}$ (4) 4

20 What is a solution for the system of equations $x - y = 2$ and $y = 2x - 4$?

(1) (0,2)   (3) (3,2)

(2) (2,0)   (4) (4,2)

21 If $a \rightarrow b$ and $c \rightarrow \sim b$ are given, which statement must be true?

(1) $a \rightarrow c$ (3) $c \rightarrow a$

(2) $b \rightarrow a$ (4) $c \rightarrow b$

22 Which equation represents the graph of a circle?

(1) $y = x$
(2) $y = x^2$
(3) $x^2 + y^2 = 9$
(4) $x = 4$

23 What is the length of the altitude of an equilateral triangle whose side has length 4?

(1) $2\sqrt{3}$ (3) $4\sqrt{3}$
(2) 2 (4) 4

24 What statement about two equilateral triangles is *always* true?

(1) They are similar.
(2) They are congruent.
(3) They are equal in the area.
(4) They have congruent altitudes.

25 In the accompanying diagram of right triangle ABC, the hypotenuse is $\overline{AB}$, $AC = 3$, $BC = 4$, and $AB = 5$.

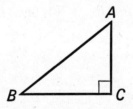

Sin B is equal to

(1) $\sin A$ (3) $\tan A$
(2) $\cos A$ (4) $\cos B$

26 If C is the midpoint of line segment $\overline{AB}$ and D is the midpoint of line segment $\overline{AC}$, which statement is true?

(1) $AC > BC$ (3) $DB = AC$
(2) $AD < CD$ (4) $DB = 3CD$

27 Which statement is logically equivalent to $\sim a \rightarrow b$?

(1) $a \rightarrow \sim b$ (3) $\sim b \rightarrow a$
(2) $b \rightarrow \sim a$ (4) $\sim b \rightarrow \sim a$

28 In the accompanying diagram of $\triangle ABC$, if $AB < BC < AC$, then which statement is *false*?

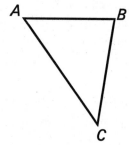

(1) $m\angle A > m\angle C$ (3) $m\angle B > m\angle C$
(2) $m\angle A < m\angle B$ (4) $m\angle B < m\angle A$

29 In the accompanying diagram of quadrilateral *ABCD*, diagonal $\overline{AC}$ bisects $\angle BAD$ and $\angle BCD$.

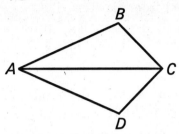

Which statement can be used to prove that $\triangle ABC \cong \triangle ADC$?

(1) HL $\cong$ HL

(2) SSS $\cong$ SSS

(3) ASA $\cong$ ASA

(4) SAS $\cong$ SAS

30 How many different six-letter arrangements can be formed using the letters in the word "DIVIDE"?

(1) 6!

(2) $_6P_6$

(3) $\dfrac{6!}{2!\,2!}$

(4) $\dfrac{6!}{4!}$

31 The roots of the equation $x^2 + 3x - 1 = 0$ are

(1) $\dfrac{-3 \pm \sqrt{5}}{2}$

(2) $\dfrac{-3 \pm \sqrt{13}}{2}$

(3) $\dfrac{3 \pm \sqrt{5}}{2}$

(4) $\dfrac{3 \pm \sqrt{13}}{2}$

32 Which is an equation of the axis of symmetry of the graph of the equation $y = x^2 - 6x + 2$?

(1) $x = -3$

(2) $y = -3$

(3) $x = 3$

(4) $y = 3$

33 In the coordinate plane, what is the total number of points that are 8 units from the origin and equidistant from the axes?

(1) 1   (3) 0
(2) 2   (4) 4

34 In rhombus *PQRS*, diagonals $\overline{PR}$ and $\overline{QS}$ intersect at *T*. Which statement is *always* true?

(1) Quadrilateral *PQRS* is a square.
(2) Triangle *RTQ* is a right triangle.
(3) Triangle *PQS* is equilateral.
(4) Diagonals $\overline{PR}$ and $\overline{QS}$ are congruent.

35 In the accompanying diagram, the bisector of an angle has been constructed.

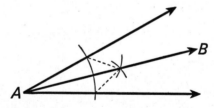

In proving this construction, which reason is used for the congruence involved?

(1) ASA   (3) AAS
(2) SSS   (4) SAS

Part Two

Answer *three* questions from this part. Clearly indicate the necessary steps, including appropriate formula substitutions, diagrams, graphs, charts, etc. Calculations that may be obtained by mental arithmetic or the calculator do not need to be shown. [30]

36 For all values of x for which these expressions are defined, perform the indicated operation and express in simplest form.

a $\dfrac{3x - 9}{x^2 - 9} - \dfrac{1}{x + 3}$ [5]

b $\dfrac{x^2 + 3x - 4}{5x - 5} \bullet \dfrac{10x^2 - 40x}{x^2 - 16}$ [5]

37 a On graph paper, draw the graph of the equation $y = -x^2 - 2x + 8$, including all values of x in the interval $-5 \leq x \leq 3$. [6]

b On the same set of axes, draw the graph of the equation $y = x + 4$. [2]

c What is the solution for the following system of equations?

$y = -x^2 - 2x + 8$
$y = x + 4$ [1,1]

38 The vertices of $\triangle ABC$ are $A(-3,-2)$, $B(2,3)$, and $C(5,-4)$.

   a On graph paper, draw and label $\triangle ABC$. [1]

   b Graph and state the coordinates of $\triangle A'B'C'$, the image of $\triangle ABC$ after a dilation of 2. [3]

   c Find the area of $\triangle A'B'C'$. [6]

39 In the accompanying diagram of isosceles trapezoid $ABCD$, $m\angle A = 53$, $DE = 6$, and $DC = 10$.

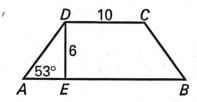

   a Find AE to the *nearest tenth*. [4]

   b Find, to the *nearest integer*, the perimeter of isosceles trapezoid $ABCD$. [6]

40 *a* Find the positive solution of $3x^2 + 2x = 7$ to the *nearest tenth*. [4]

  b Given: If I receive a check for \$500, then we will go on a trip.

   If the car breaks down, then we will not go on a trip.

   Either I receive a check for \$500 or we will not buy souvenirs.

   The car breaks down.

Get *C* represent: "I receive a \$500 check."
Let *T* represent: "We will go on a trip."
Let *B* represent: "The car breaks down."
Let *S* represent: "We will buy souvenirs."

Using the laws of logic, prove that we will not buy souvenirs. [6]

Part III

Answer *one* question from this part. Clearly indicate the necessary steps, including appropriate formula substitutions, diagrams, graphs, charts, etc. Calculations that may be obtained by mental arithmetic or the calculator do not need to be shown. [10]

41 Given: rectangle $ABCD$, $\overline{BNPC}$, $\overline{AEP}$, $\overline{DEN}$, and $\overline{AP} \cong \overline{DN}$.

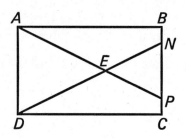

Prove: a $\triangle ABP \cong \triangle DCN$ [5]

   b $\overline{AE} \cong \overline{DE}$ [5]

42 The vertices of quadrilateral $GAME$ are $G(r,s)$, $A(0,0)$, $M(t,0)$, and $E(t + r,s)$. Using coordinate geometry, prove that quadrilateral $GAME$ is a parallelogram. [10]

ANSWER KEY

Part I

1. 4
2. 130
3. $\overline{PE}$
4. 8
5. 10
6. 5
7. −1
8. (−5, −3)
9. 70
10. 10
11. 90
12. 110

13. (1, −10)
14. 35
15. 80
16. $3(x + 2)(x − 7)$
17. (1)
18. (1)
19. (4)
20. (2)
21. (1)
22. (3)
23. (1)
24. (1)

25. (2)
26. (4)
27. (3)
28. (4)
29. (3)
30. (3)
31. (2)
32. (3)
33. (4)
34. (2)
35. (2)

EXPLANATIONS: JANUARY 1996

Part I

1. 4

This function question defines what the "Δ" means, so all you have to do is plug in $a = 2$ and $b = 4$:

$$a\Delta b = \frac{2a + b}{a}$$

$$2\Delta 4 = \frac{2(2) + 4}{2} = \frac{4 + 4}{2} = \frac{8}{2} = 4$$

2. 130

You can use a few geometric theorems to solve this one. Since the problem tells you that $\overline{AB} \cong \overline{AC}$, it must also be true that the angles opposite these two sides, $\angle ACB$ and $\angle B$, are congruent. Set each of these angles equal to x, and use the Rule of 180:

$$\text{m}\angle A + \text{m}\angle B + \text{m}\angle ACB = 180$$
$$80 + x + x = 180$$
$$2x = 100$$
$$x = 50$$

Now you know that $\angle B$ and $\angle ACB$ measure 50° each, but you're not done yet. $\angle ACD$ is an exterior angle, and the measure of any exterior angle of a triangle is equal to the sum of the two non-adjacent interior angles. Therefore:

$$\text{m}\angle ACD = \text{m}\angle A + \text{m}\angle B = 80 + 50 = 130$$

Note: You can also find the measure of $\angle ACD$ if you recognize that $\angle ACD$ and $\angle ACB$ are supplementary.

3. $\overline{PE}$

When the test doesn't give you a diagram, it helps to draw one on your own.

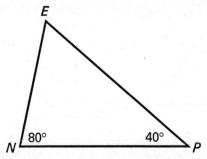

Using the Rule of 180, you know that m $\angle E$ = 60. In any triangle, the longest side is always opposite the largest angle. Since $\angle N$ is the biggest angle, the side opposite that angle, $\overline{PE}$, must be the biggest side.

4. 8

Whenever two fractions equal each other, you can cross-multiply:

$$\frac{x-2}{x+4} = \frac{x+2}{x+12}$$
$$(x-2)(x+12) = (x+4)(x+2)$$

Now, use FOIL to simplify each side of the equation:

$$x^2 + 12x - 2x - 24 = x^2 + 4x + 2x + 8$$

Since x^2 appears on each side of the equation, you can cross them out. Once you combine the like terms, you should come up with this:

$$10x - 24 = 6x + 8$$
$$4x - 24 = 8$$
$$4x = 32$$
$$x = 8$$

To check your work, plug 8 back into the original equation and make sure it works.

5. 10

Any time a problem involves similar triangles, all you do is set up a proportion. The key is lining up the corresponding sides. First draw a diagram:

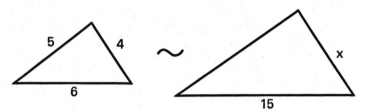

From the diagram, you can see that the longest side of the smaller triangle is 6 units long, and its counterpart of the larger triangle is 15 units long. The shortest side of the smaller triangle is 4 units long, and you're looking for the length of its corresponding side in the larger triangle; call it x. Your proportion should look like this:

$$\frac{6}{15} = \frac{4}{x}$$

Cross-multiply and solve:

$$6x = 60$$
$$x = 10$$

6. 5

Whenever a question involves plotting points, draw as accurate a diagram as you can. It should look like this:

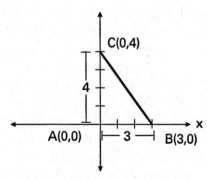

Once you look at the diagram, you should recognize △*ABC* as a right triangle. Since *AC* = 4 and *AB* = 3, you've got a 3:4:5 right triangle on your hands. The hypotenuse, $\overline{BC}$, must therefore be 5 units long.

The Regents folks want you to get bogged down using the distance formula or the Pythagorean Theorem. It's OK to use them if you have to, but it's always a good idea to look for ways to save time.

7. –1

Use the slope formula:

$$m = \frac{y_2 - y_1}{x_2 - x_1}$$

Let $(x_1, y_1) = (-1, 3)$ and $(x_2, y_2) = (3, -1)$:

$$m = \frac{-1 - 3}{3 - (-1)} = \frac{-4}{4} = -1$$

8. (–5, –3)

Whenever you reflect a point in the *y*-axis, the *x*-coordinate is negated and the *y*-coordinate remains the same. The formula for such a reflection looks like this:

$$r_{y\text{-}axis}(x, y) = (-x, y)$$

Therefore, the image of the point *N*(5, –3) is *N'*(–5, –3).

9. 70

From the diagram, you can tell that ∠1 and ∠2 are supplementary. Set up an equation indicating that their sum equals 180 and solve for *x*:

$$(2x + 20) + (4x + 10) = 180$$
$$6x + 30 = 180$$
$$6x = 150$$
$$x = 25$$

Hold it. You're not done. You have to substitute 25 for *x* to figure out the exact measure of each angle: m∠1 = (2 × 25) + 20, or 70, and m∠2 = (4 × 25) + 10, or 110. (Since 70 + 110 = 180, you're doing OK so far.)

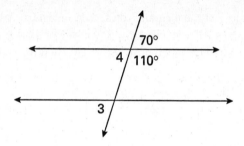

Since ∠4 and ∠1 are vertical angles (which have the same measure), m∠4 = 70. Since ∠3 and ∠4 are corresponding angles, they also have the same measure. Therefore, m∠3 = 70.

10. 10

First, plot the triangle on the coordinate axes:

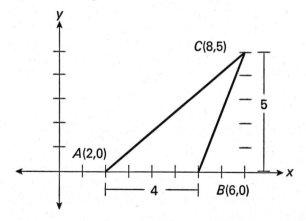

The formula for the area of a triangle is $\frac{1}{2}bh$. The base of the triangle, $\overline{AB}$, is 4 units long. The height is a little trickier; you have to drop a perpendicular from point C to the x-axis, which contains $\overline{AB}$. That distance, 5, represents the height of the triangle. Once you figure out those two lengths, you can plug them in to the formula:

$$A = \frac{1}{2}(4)(5) = 10$$

11. 90

First, draw a triangle, and label two of the angles a and b. Since the third angle is the sum of the other two, its measure must equal $a + b$:

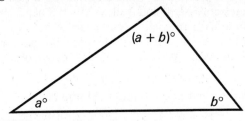

Use the Rule of 180; the sum of all three angles equals 180:

$$a + b + (a + b) = 180$$
$$2a + 2b = 180$$
$$2(a + b) = 180$$
$$a + b = 90$$

The largest angle must measure 90° (it's a right angle).

12. 110

This one looks like a real stinker at first glance, but it becomes rather simple if you add a line to the diagram. Draw a line parallel to both AB and CD through the vertex of $\angle z$ and call it $\overline{MN}$, like so:

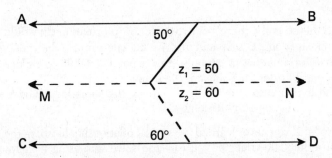

As you can see, this line cuts $\angle z$ into two angles: $\angle z_1$ and $\angle z_2$. Look closely at $\angle x$ and $\angle z_1$: they're alternate interior angles! Therefore, they must have the same measure, 50°. Similarly, $\angle y$ and $\angle z_2$ are also alternate interior angles. Thus, m$\angle z_2$ = 60. Now solve for m$\angle z$:

$$\text{m}\angle z = \text{m}\angle z_1 + \text{m}\angle z_2 = 60 + 50 = 110$$

13. (1,–10)

The easiest way to figure out a translation is to subtract the coordinates of the original point from those of the image. Look at the x-coordinates of $A(-3, 4)$ and $A'(2, -6)$ first. Since $2 - (-3) = 5$, the translation adds 5 to each x-coordinate. The difference of the y-coordinates $(-6 - 4 = -10)$ tells you that this same translation subtracts 10 from each y-coordinate. The formula looks like this:

$$(x, y) \rightarrow (x + 5, y - 10)$$

Now, plug the point $B(-4, 0)$ into the formula:

$$B' = (-4 + 5, 0 - 10) = (1, -10)$$

14. 35

This is a combinations problem (because the order of the students on the committee doesn't matter), so use the combinations formula:

$$_nC_r = \frac{n!}{r!(n - r)!}$$

There are seven students ($n = 7$), and you want to choose three of them ($r = 3$):

$$_7C_3 = \frac{7!}{3!\,4!} = \frac{7 \times 6 \times 5 \times 4 \times 3 \times 2 \times 1}{(3 \times 2 \times 1) \times (4 \times 3 \times 2 \times 1)} = \frac{210}{6} = 35$$

15. 80

In this triangle, there's one extra bit of information they didn't tell you: since m $\angle C$ = 90 and m $\angle A$ = 60, you can use the Rule of 180 to determine that m $\angle B$ = 30. You've got a 30:60:90 triangle! Side b is the short leg, and side c is the hypotenuse. In a 30:60:90 triangle, the hypotenuse is twice the length of the short leg; since $b = 40$, side c must be 80 units long.

Note: If you never learned the special relationship between the sides of a 30:60:90 triangle, you can also solve this using trigonometry. You know that m $\angle A$ = 60, and the length of the adjacent leg is 40. You're looking for the length of the hypotenuse, so use cosine (the CAH in SOHCAHTOA):

$$\cos 60° = \frac{40}{c}$$

$$0.5 = \frac{40}{c}$$

$$c = \frac{40}{0.5}$$

$$c = 80$$

16. $3(x + 2)(x - 7)$

Factor 3 out of each term:

$$3x^2 - 15x - 42 = 3(x^2 - 5x - 14)$$

Now factor the binomial in the parentheses:

$$x^2 - 5x - 14 = (x + 2)(x - 7)$$

The entire factorization becomes:

$$3(x + 2)(x - 7)$$

Once you're done, try FOILing the terms back together again to make sure you're correct.

Multiple Choice

17. (1)

De Morgan's Laws look like this:

$$\sim(a \wedge b) \rightarrow \sim a \vee \sim b; \ \sim(a \vee b) \rightarrow \sim a \wedge \sim b$$

This basically means than when you negate a parenthetical statement with a "$\wedge$" or "$\vee$" in it, negate each symbol and turn the symbol upside down:

$$\sim(p \wedge \sim q) \rightarrow \sim p \vee \sim(\sim q)$$

Since $\sim(\sim q)$ is the same thing as q (because of the rule of double negation), you can rewrite the statement as: $\sim p \vee q$.

18. (1)

The additive inverse states that if the sum of any two numbers is zero, one of the numbers is the opposite of the other one. (That is, –3 is the additive inverse of 3.) Answer choice (2) is the additive *identity*, not the inverse, so cross is out. Eliminate answer choices (3) and (4); they don't even involve addition!

19. (4)

First draw your diagram:

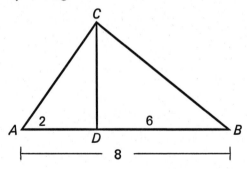

This figure represents three similar right triangles ($\triangle ADC$, $\triangle CDB$, and $\triangle ACB$), and all of their corresponding sides are proportional to each other. For this problem, consider $\triangle ADC$ and $\triangle ACB$ and set up a proportion. In $\triangle ACB$, AB is the hypotenuse and AC is the short leg; in $\triangle ADC$, AC is the hypotenuse and AD is the short leg. Therefore:

$$\frac{AB}{AC} = \frac{AC}{AD}$$

$$\frac{8}{AC} = \frac{AC}{2}$$

When you cross-multiply, you'll get:

$$(AC)^2 = 16$$
$$AC = 4, -4$$

Since you're looking for a distance, which has to be positive, forget about –4.

20. (2)

Problems like this one cry out to be backsolved. Simply plug the values of x and y into the two equations and see which coordinate pair works in both of them. Answer choices (1) and (3) don't fit into $x - y = 2$, and answer choice (4) doesn't work for $y = 2x - 4$. Only answer choice (2) works in both of them:

$$x - y = 2 \qquad\qquad y = 2x - 4$$

$$2 - 0 = 2 \qquad\qquad 0 = 2(2) - 4$$

Note: Be sure not to get your coordinates confused: the first number in a coordinate pair is the x-coordinate.

21. (1)

The key to this logic question is the manipulation of the second statement. Using the Law of Contrapositive Inference (otherwise known as the Flip-and-Negate Rule), $\sim c \to \sim b$ can be rewritten as $b \to c$. Now it's time for the Chain Rule (or Syllogism): if $a \to b$ and $b \to c$ then it can be inferred that $a \to c$.

22. (3)

Use Process of Elimination on this one. You know that answer choices (1) and (4) are lines and answer choice (2) is a parabola, so eliminate them. Remember also that the formula for a circle is $x^2 + y^2 = r^2$, and r represents the radius of the circle. In this case, the radius is 3.

23. (1)

First, draw the triangle:

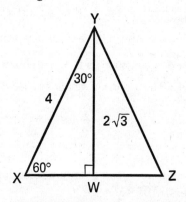

As you can see, the altitude has to be smaller than a side of the triangle. Therefore, you can eliminate answer choices (3) and (4) because they're too big. Now, look closely at $\triangle XYW$. Altitude $\overline{YW}$ cuts the equilateral triangle into two 30:60:90 triangles. If side $\overline{XY}$ has a length of 4, then $XW = 2$ and $YW = 2\sqrt{3}$.

24. (1)

Process of Elimination works great here. You can cross off answer choices (2), (3), and (4) right away, because none of those answers is true unless both triangles are the same size. All equilateral triangles are the same shape, so they must be similar.

25. (2)

You're going to use SOHCAHTOA here. First, write in all the lengths the question dictates, like so:

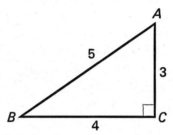

Sin B equals $\dfrac{\text{opposite}}{\text{hypotenuse}}$, or $\dfrac{3}{5}$. Now, it's just a matter of substituting the values of the answer choices until you find a match. Since $\sin A = \dfrac{4}{5}$, $\tan A = \dfrac{4}{3}$, and $\cos B = \dfrac{4}{5}$, eliminate answer choices (1), (3), and (4). The right answer is answer choice (2), because $\cos A = \dfrac{3}{5}$.

26. (4)

Draw your diagram first, and plug in a value for the length of the smallest segment, $\overline{AD}$. Say $AD = 2$:

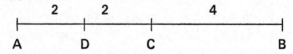

Now eliminate: AC and BC are the same size, so cross off answer choice (1). Ditto for answer choice (2). You can get rid of answer choice (3) because $AC = 4$ and $DB = 6$. Answer choice (4) works, because $CD = 2$.

27. (3)

It's time once again for the Flip-and-Negate Rule (the Law of Contrapositive Inference). When you flip and negate the statement ~$a \rightarrow b$, you get ~$b \rightarrow a$.

28. (4)

Here's another test of the relationship between the sides and angles of a triangle. Since $\overline{AB}$ is the smallest side, the angle opposite it, $\angle C$, is the smallest angle. Using similar logic, you can determine that $\angle B$ is the triangle's biggest angle. From this information, you can deduce the following:

$$\text{m}\angle C < \text{m}\angle A < \text{m}\angle B$$

Answer choice (4) is the only one that violates this rule.

29. (3)

First of all, cross off (1). Hypotenuse-Leg only works with right triangles. Now, take a look at what you know. Since diagonal $\overline{AB}$ bisects $\angle BAD$, you know that $\angle BAC \cong \angle CAD$. Similarly, you can deduce that $\angle BCA \cong \angle ACD$. The right answer has to involve two angles (which means it has to have two A's in it), and answer choice (3) is the only one that does. Just to be thorough, check out the diagram:

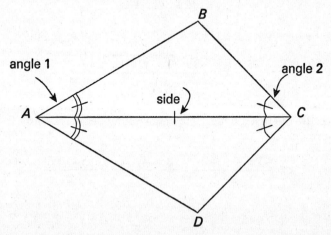

Using the Reflexive Property, you can deduce that $\overline{AC} \cong \overline{AC}$. That's the side in Angle-Side-Angle.

30. (3)

Answer choice (1) would be correct if there were no repeating letters. Since we have some repetition, however, eliminate it. Cross off answer choice (2) also, because $_6P_6$ is the same thing.

The formula to follow is a variation of the permutations rule. To find the number of possible arrangements of the letters in a word with n letters, in which one letter appears p times and another letter appears q times (remember that both p and q are greater than 1), the formula looks like this:

$$\frac{n!}{p!\,q!}$$

DIVIDE has six letters, but there are two D's and two I's. Therefore, you can express the number of arrangements as:

$$\frac{6!}{2!\,2!}$$

31. (2)

Since all the answer choices have irrational roots, you know that you can't factor the equation without using the Quadratic Formula, which looks like this:

$$x = \frac{-b \pm \sqrt{b^2 - 4ac}}{2a}$$

In the equation , $a = 1$, $b = 3$, and $c = -1$. So plug 'em in:

$$x = \frac{-3 \pm \sqrt{3^2 - 4(1)(-1)}}{2(1)}$$

Note: At this point, you can eliminate answer choices (3) and (4) because the first number in the numerator of each is *positive* 3. Let's move on:

$$x = \frac{-3 \pm \sqrt{9 - (-4)}}{2} = \frac{-3 \pm \sqrt{9 + 4}}{2} = \frac{-3 \pm \sqrt{13}}{2}$$

32. (3)

When a parabola appears in the form $y = ax^2 + bx + c$ the equation of its axis of symmetry is:

$$x = -\frac{b}{2a}$$

For this parabola, $a = 1$, $b = -6$, and $c = 2$. Thus the axis of symmetry is:

$$x = -\frac{-6}{2(1)} = \frac{6}{2} = 3$$

The formula for the line is $x = 3$.

Note: You could also have eliminated answer choices (2) and (4). Since this parabola goes up and down, the axis of symmetry is vertical and involves x, not y.

33. (4)

First, consider the locus of points that are 8 units from the origin. It's a circle with radius 8:

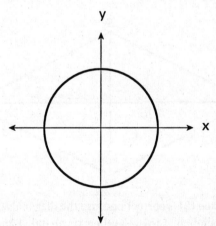

Now, think about the set of points that are equidistant from the coordinate axes. The drawing will look like two lines that pass through the origin and form a 45-degree angle with each axis, like this:

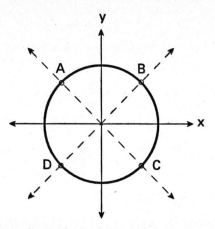

These two sets of points intersect at the four points indicated above. There are four points that satisfy both criteria.

34. (2)

You can eliminate answer choice (1) right away, because not every rhombus is a square. Now, draw a diagram, and be sure to draw a rhombus that is NOT a square:

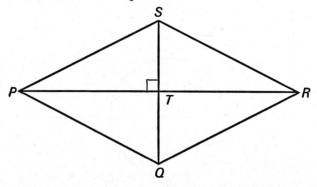

Answer choice (2) is correct because the diagonals of a rhombus are always perpendicular to each other. If you didn't know that, you can still get the right answer if you use your diagram and the Process of Elimination. Does △PQS look equilateral? Nope. Eliminate answer choice (3). Do diagonals $\overline{PR}$ and $\overline{QS}$ appear to be the same length? No. Get rid of answer choice (4).

35. (2)

This is a tricky one. How often are you asked to analyze a construction problem? After a little time to think about it, it might come to you. Consider the diagram below (we've labeled a few of the important points):

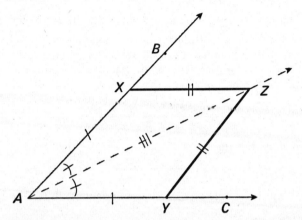

With your first swipe with the compass, points X and Y represent where the arc you draw intersects rays AB and AC, respectively. Therefore, $\overline{AX} \cong \overline{AY}$. The arcs drawn from points X and Y that intersect at point Z are drawn with the same compass setting, so $\overline{XZ} \cong \overline{YZ}$. Using the reflexive property, you know that $\overline{AZ} \cong \overline{AZ}$. Now you can prove that $\triangle AXZ \cong \triangle AYZ$ using SSS. From there, you can prove that $\angle XAZ \cong YAZ$ because corresponding parts of congruent triangles are congruent (CPCTC).

Part II

36. ***a*** $\dfrac{2}{x+3}$

Before you combine fractions, they must have the same denominator. The key to this problem is simplifying the term $\dfrac{3x-9}{x^2-9}$. You can factor 3 out of both terms in the numerator; now it looks like $3(x-3)$. The denominator is a difference of squares. Using the formula $x^2 - y^2 = (x+y)(x-y)$, you can factor $x^2 - 9$ into $(x+3)(x-3)$. The problem now looks like this:

$$\frac{3(x-3)}{(x+3)(x-3)} - \frac{1}{x+3}$$

which simplifies to:

$$\frac{3}{x+3} - \frac{1}{x+3}$$

Now it's just a simple case of subtracting numerators. The final answer is $\dfrac{2}{x+3}$.

b $2x$

It's hard to believe that something so complicated can reduce to something so simple, but it's possible. It's time to showcase all your factoring skills:

$$x^2 + 3x - 4 = (x+4)(x-1)$$
$$5x - 5 = 5(x-1)$$
$$10x^2 - 40x = 10x(x-4)$$
$$x^2 - 16 = (x+4)(x-4)$$

Once you've factored these four terms, the problem looks like this:

$$\frac{(x+4)(x-1)}{5(x-1)} \cdot \frac{10x(x-4)}{(x+4)(x-4)}$$

Now for the fun part: cancellation. If you cancel out all the factors that appear both on the top and on the bottom, you're left with:

$$\frac{\cancel{(x+4)}\cancel{(x-1)}}{5\cancel{(x-1)}} \cdot \frac{10x\cancel{(x-4)}}{\cancel{(x+4)}\cancel{(x-4)}} = \frac{10x}{5}$$

This last expression simplifies down to $2x$.

37. *a*

You know the interval for *x* you've been given, so figure out your T-chart by substituting the *x*-values into the equation, like so:

If $x = -5$, then: $y = -(-5)^2 - 2(-5) + 8 = -7$. Your first ordered pair is: $(-5, -7)$.

Here's the rest of the T-chart, along with the sketch of the parabola:

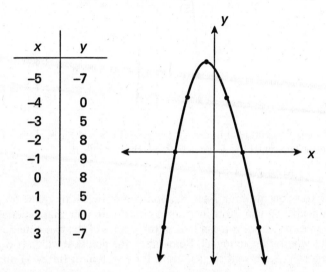

x	y
–5	–7
–4	0
–3	5
–2	8
–1	9
0	8
1	5
2	0
3	–7

b

Drawing the line $y = x + 4$ is a snap. All you have to do is plot the *x*- and *y*-intercepts and connect them:

If $x = 0$, then $y = 0 + 4$, or 4. Plot (0, 4) on the graph.

If $y = 0$, then $0 = x + 4$, and $x = -4$. Plot (–4, 0) on the graph, and connect them like this:

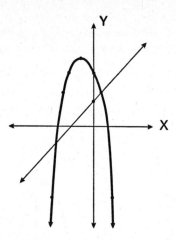

Note: You can also note that the line $y = x + 4$ has a slope of 1 and intersects with the y-axis at $(0, 4)$, then graph it.

c **(1,5) and (–4,0)**

If you consult your T-charts, you can answer this question rather quickly. All you do is look for a coordinate pair that satisfies both equations. Also, you can look at the graphs of the two equations and see where they intersect. Remember, the points at which two graphs intersect represent the solutions for a system of the two equations

Note: As a sophisticated way to double-check your work, solve the system algebraically. Since $y = -x^2 - 2x + 8$ and $y = x + 4$, you can set the two values of y equal to each other:

$$-x^2 - 2x + 8 = x + 4$$
$$-x^2 - 3x + 4 = 0$$

Multiply through by –1 to make the equation easier to factor:

$$x^2 + 3x - 4 = 0$$
$$(x + 4)(x - 1) = 0$$
$$x = \{-4, 1\}$$

Now you have the two values of x. As you've calculated before, if $x = -4$ then $y = 0$; if $x = 1$, then $y = 5$. The two ordered pairs are $(-4,0)$ and $(1,5)$.

38. *a*

This first part is easy. No wonder it was worth only 1 point!

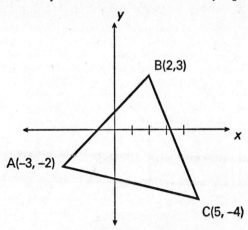

b

When a point undergoes a dilation of 2, each coordinate of the point is multiplied by 2. Therefore, the image of $A(-3,-2)$ is $A'(-6,-4)$; the image of $B(2,3)$ is $B'(4,6)$; and the image of $C'(5,-4)$ is $C'(10,-8)$. Your graph of $\triangle A'B'C'$ looks like this:

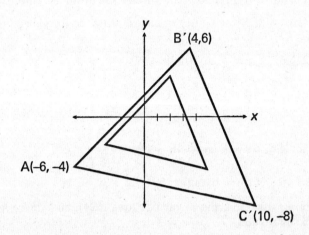

c 100

The best way to determine the area of $\triangle A'B'C'$ is to work indirectly. First, circumscribe a rectangle on the triangle in the diagram, like so:

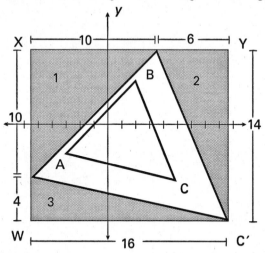

Triangle $A'B'C'$ is now surrounded by three other triangles within the rectangle. If you subtract their combined area from the area of the rectangle, the result will be the area of $\triangle A'B'C'$. The area of rectangle $WXYC'$ is 16 × 14, or 224 square units. Now, calculate the area of the three outer triangles, using the formula $A = \frac{1}{2}bh$

Triangle 1: $A = \frac{1}{2}bh\,(10)(10) = 50$

Triangle 2: $A = \frac{1}{2}bh\,(6)(14) = 42$

Triangle 3: $A = \frac{1}{2}bh\,(4)(16) = 32$

Total area of three triangles: 50 + 42 + 32 = 124

Subtract 124 from the rectangle's area (224), and you're left with **100** square units

39. *a* 4.5

To figure out the length of $\overline{AE}$, you have to use a little trigonometry. You know the measure of $\angle A$ and the length of $\overline{DE}$, so use the tangent (the TOA in SOHCAHTOA):

$$\tan \angle A = \frac{DE}{AE}$$

$$\tan 53° = \frac{6}{AE}$$

$$1.3270 = \frac{6}{AE}$$

$$1.3270(AE) = 6$$

$$AE = \frac{6}{1.3270} = 4.5$$

***b* 44**

On the diagram, drop a perpendicular from C to $\overline{AB}$, and label the point of intersection H:

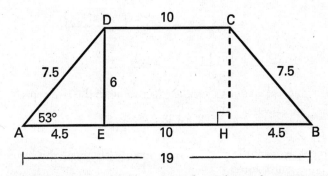

Since $ABCD$ is an isosceles trapezoid, you know that $AE = BH = 4.5$. Therefore, the base of the trapezoid equals $4.5 + 10 + 4.5$, or 19. You also know that $CD = 10$, so there's one missing piece of information: the lengths of $\overline{AD}$ and $\overline{BC}$. This is an isosceles trapezoid, so they're equal. Use the sine (the SOH in SOHCAHTOA) to figure out AD:

$$\sin \angle A = \frac{6}{AD}$$

$$\sin 53° = \frac{6}{AD}$$

$$0.7986 = \frac{6}{AD}$$

$$0.7986(AD) = 6$$

$$AD = \frac{6}{0.7986} \approx 7.5$$

Now you know that $AD = BC = 7.5$, and you can add up the sides:

$$7.5 + 19 + 7.5 + 10 = 44$$

40. *a* 1.2

This looks like a job for the Quadratic Formula. First, you have to put the quadratic in standard form (which means you have to set everything equal to zero) like this:

$$3x^2 + 2x = 7$$
$$3x^2 + 2x - 7 = 0$$

Now use the formula, noting that $a = 3$, $b = 2$, and $c = -7$:

$$x = \frac{-b \pm \sqrt{b^2 - 4ac}}{2a} = \frac{-2 \pm \sqrt{2^2 - 4(3)(-7)}}{2(3)}$$

$$= \frac{-2 \pm \sqrt{4 + 84}}{6} = \frac{-2 \pm \sqrt{88}}{6}$$

Use your calculator to get the exact value of the two roots:

$$x = \frac{-2 + 9.38}{6}, \frac{-2 - 9.38}{6}$$

Since you want the positive root, you can eliminate the second possibility. Solve for x:

$$x_1 = \frac{7.38}{6} = 1.2$$

b

In this proof, they've already given you the symbols.

Step 1: Turn all the givens into symbolic terms:

"If I receive a check for \$500, then we will go on a trip." $C \to T$

"If the car breaks down, then we will not go on the trip." $B \to \sim T$

"Either I receive a check for \$500 or we will not buy souvenirs." $C \lor \sim S$

"The car breaks down." B

Step 2: Decide what you want to prove:

"We will not buy souvenirs." $\sim S$

Step 3: Write the proof.

Statements	Reasons
1. B	1. Given
2. $B \to \sim T$	2. Given
3. $\sim T$	3. Law of Detachment (1,2)
4. $C \to T$	4. Given
5. $\sim C$	5. Law of *Modus Tollens* (3,4)
6. $C \lor \sim S$	6. Given
7. $\sim S$	7. Law of Disjunctive Inference (5,6)

(**Note:** If your class didn't cover the Law of *Modus Tollens*, you can use a combination of the Law of Contrapositive Inference and the Law of Detachment.)

Part III

41. *a*

The plan: Since △*ABP* and △*DCN* are right triangles, think Hypotenuse-Leg.

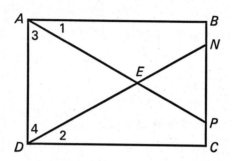

Statements	Reasons
1. *ABCD* is a rectangle.	1. Given.
2. ∠*A*, ∠*B*, ∠*C*, and ∠*D* are right angles.	2. A rectangle contains four right angles.
3. △*ABP* and △*DCN* are right triangles.	3. Definition of a right triangle.
4. $\overline{AP} \cong \overline{DN}$	4. Given
5. $\overline{AB} \cong \overline{CD}$	5. Opposite sides of a rectangle are congruent.
6. △*ABP* ≅ △*DCN*	6. HL ≅ HL

b The plan: Prove that $\triangle AED$ is isosceles.

Statements	Reasons
1. $\triangle ABP \cong \triangle DCN$	1. Given (from previous proof)
2. $m\angle 1 = m\angle 2$	2. Corresponding parts of congruent triangles are congruent (CPCTC).
3. $m\angle BAD = m\angle CDA$	3. All right angles have the same measure.
4. $m\angle BAD = m\angle 1 + m\angle 3$; $m\angle CDA = m\angle 2 + m\angle 4$	4. Angle Addition Postulate
5. $m\angle 3 = m\angle 4$	5. Subtraction Property of Equality
6. $\overline{AE} \cong \overline{DE}$	6. If two angles of a triangle are equal in measure, the sides opposite them are congruent.

42.

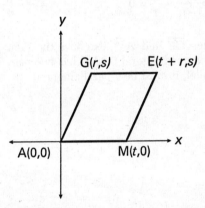

Don't be scared off by the variables in the coordinates. Just use the variables as you would regular numbers. Prove that opposite sides of *GAME* are parallel using the slope formula, which looks like this:

$$m = \frac{y_2 - y_1}{x_2 - x_1}$$

First, find the slopes of opposite sides $\overline{GA}$ and $\overline{ME}$:

Slope of $\overline{GA}$:

$(x_1, y_1) = G(r, s);$

$(x_2, y_2) = A(0, 0);$

$m = \dfrac{0 - s}{0 - r} = \dfrac{-s}{-r} = \dfrac{s}{r}$

Slope of $\overline{ME}$:

$(x_1, y_1) = M(t, 0);$

$(x_2, y_2) = E(t + r, s);$

$m = \dfrac{s - 0}{(t + r) - t} = \dfrac{s}{t + r - t} = \dfrac{s}{r}$

Opposite sides $\overline{GE}$ and $\overline{AM}$ have the same slope, so they're parallel. Now find the slopes of opposite $\overline{GE}$ sides and $\overline{AM}$:

Slope of $\overline{GE}$:

$(x_1, y_1) = G(r, s);$

$(x_2, y_2) = E(t + r, s)$

$m = \dfrac{s - s}{(t + r) - r} = \dfrac{0}{t} = 0$

Slope of $\overline{AM}$:

$(x_1, y_1) = A(0, 0);$

$(x_2, y_2) = M(t, 0)$

$m = \dfrac{0 - 0}{t - 0} = \dfrac{0}{t} = 0$

Opposite sides $\overline{GE}$ and $\overline{AM}$ also have the same slope, so they're parallel. Since quadrilateral $GAME$ has two pairs of opposite sides that are parallel, it must be a parallelogram.

EXAMINATION:
JUNE 1996

Part I

Answer 30 questions from this part. Each correct answer will receive 2 credits. No partial credit will be allowed. Write your answers in the spaces provided on the separate answer sheet. Where applicable, answers may be left in terms of π or in radical form. [60]

1 If $a \odot b$ is defined as $a - 2b$, find the value of $5 \odot 7$.

2 If $\tan A = 1.3400$, find the measure of $\angle A$ to the *nearest degree*.

3 What is the identity element in the system defined by the table below?

$\star$	2	4	6	8
2	4	8	2	6
4	8	6	4	2
6	2	4	6	8
8	6	2	8	4

4 In the accompanying figure, $\overline{DE} \parallel \overline{BC}$, $AD = 10$, $AB = 24$, and $AC = 36$. Find AE.

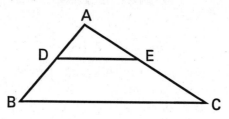

5 Evaluate: $_7C_3$

6 If one the roots of the equation $x^2 + kx = 6$ is 2, what is the value of k?

7 Solve for the positive value of y: $\dfrac{16}{y} = \dfrac{y}{4}$

8 How many different 4-letter arrangements can be formed from the letters in the word "NINE"?

9 In $\triangle ABC$, $m \angle B > m \angle C$ and $m \angle C > m \angle A$. Which side of $\triangle ABC$ is the longest?

10 In the accompanying diagram of rhombus $ABCD$, diagonal $\overline{AC}$ is drawn. If $m \angle CAB = 35$, find $m \angle ADC$.

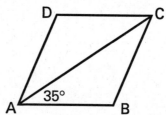

11 What is the slope of the line whose equation is $3x + y = 4$?

12 The graph of the equation $x^2 + y^2 = 9$ represents the locus of points at a given distance, d, from the origin. Find the value of d.

13 Find the area of the parallelogram whose vertices are (2,1), (7,1), (9,5), and (4,5).

14 Express $\dfrac{5x}{6} - \dfrac{x}{3}$ in simplest form.

15 The line that passes through points (1,3) and (2,y) has a slope of 2. What is the value of y?

16 What is the length of a side of a square whose diagonal measures $4\sqrt{2}$?

Directions (17-35): For *each* question chosen, write the *numeral* preceding the word or expression that best completes the statement or answers the question.

17 When factored completely, $x^3 - 9x$ is equivalent to
(1) $x(x - 3)$
(2) $x(x + 3)(x - 3)$
(3) $(x + 3)(x - 3)$
(4) $x(x + 3)$

18 If $(x + 2)^2 + (y - 3)^2 = 25$ is an equation of a circle whose center is $(-2, k)$, then k equals

(1) 1   (3) 3
(2) 2   (4) 4

19 In the accompanying diagram of $\triangle ABC$, side $\overline{BC}$ is extended to D, m$\angle B = 2y$, m$\angle BCA = 6y$, and m$\angle ACD = 3y$.

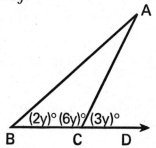

What is m$\angle A$?

(1) 15   (3) 20
(2) 17   (4) 24

20 The coordinates of $\triangle ABC$ are $A(0,0)$, $B(6,0)$, and $C(0,4)$. What are the coordinates of the point at which the median from vertex A intersects side $\overline{BC}$?

(1) (1,4)   (3) (3,0)
(2) (2,3)   (4) (3,2)

21 Which statement is the equivalent of $\sim(\sim m \wedge n)$?

(1) $m \wedge n$ (3) $m \vee \sim n$
(2) $m \wedge \sim n$ (4) $\sim m \vee \sim n$

22 The translation $(x,y) \rightarrow (x-2, y+3)$ maps the point (7,2) onto the point whose coordinates are

(1) (9,5)   (3) (5,−1)
(2) (5,5)   (4) (−14,6)

23 In the accompanying diagram, $\triangle FUN$ is a right triangle, $\overline{UR}$ is the altitude to hypotenuse $\overline{FN}$, UR = 12, and the lengths of $\overline{FR}$ and $\overline{RN}$ are in the ratio 1:9.

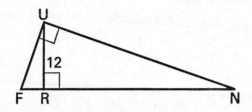

What is the length of $\overline{FR}$?

(1) 1   (3) 36

(2) $1\frac{1}{3}$ (4) 4

24 Lines l and m are perpendicular. The slope of l is $\frac{3}{5}$. What is the slope of m?

(1) $-\frac{3}{5}$ (3) $\frac{3}{5}$

(2) $-\frac{5}{3}$ (4) $\frac{5}{3}$

25 In the accompanying diagram of △ABC, $\overline{AC}$ is extended to D, $\overline{DEF}$, $\overline{BEC}$, $\overline{AFB}$, m∠B = 50, m∠BEF = 25, and m∠ACB = 65.

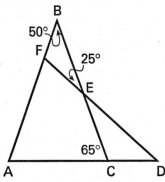

What is m∠D?

(1) 40   (3) 50
(2) 45   (4) 55

26 In the accompanying diagram, parallel lines *l* and *m* are cut by transversal *t*.

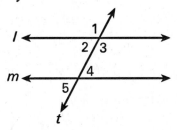

Which statement is true?

(1)   m∠1 + m∠2 + m∠5 = 360

(2)   m∠1 + m∠2 + m∠3 = 180

(3)   m∠1 + m∠2 = m∠2 + m∠3

(4)   m∠1 + m∠3 = m∠4 + m∠5

27 Which argument below is *not* valid?

   (1) Given: $a \rightarrow b$
                a
      Conclusion: b

   (3) Given: $a \rightarrow b$
                $\sim b$
      Conclusion: $\sim a$

   (2) Given: $a \vee b$
                $\sim b$
      Conclusion: $\sim a$

   (4) Given: $a \rightarrow b$
                $b \rightarrow \sim c$
      Conclusion: $a \rightarrow \sim c$

28 The measure of a base angle of an isosceles triangle is four times the measure of a vertex angle. The number of degrees in the vertex angle is

   (1) 20
   (2) 30

   (3) 36
   (4) 135

29 What are the coordinates of R', the image of $R(-4,3)$ after a reflection in the line whose equation is $y = x$?

   (1) $(-4,-3)$
   (2) $(3,-4)$

   (3) $(4,3)$
   (4) $(-3,4)$

30 The equation $y = 4$ represents the locus of points that are equidistant from which two points?

   (1) $(0,0)$ and $(0,8)$
   (2) $(0,3)$ and $(0,1)$

   (3) $(4,0)$ and $(0,4)$
   (4) $(4,4)$ and $(-4,4)$

31 In equilateral triangle ABC, $\overline{AD}$ is drawn to $\overline{BC}$ such that $BD < DC$. Which inequality is true?

   (1) $DC > AC$
   (2) $BD > AD$

   (3) $AD > AB$
   (4) $AC > AD$

32 Which equation represents the axis of symmetry of the graph of the equation $y = 2x^2 + 7x - 5$?

(1) $x = -\dfrac{5}{4}$ (3) $x = \dfrac{7}{4}$

(2) $x = \dfrac{5}{4}$ (4) $x = -\dfrac{7}{4}$

33 How many congruent triangles are formed by connecting the midpoints of the three sides of a scalene triangle?

(1) 1   (3) 3

(2) 2   (4) 4

34 What are the roots of the equation $2x^2 - 7x + 4 = 0$?

(1) $\dfrac{7 \pm \sqrt{17}}{4}$ (3) $4, \ -\dfrac{1}{2}$

(2) $\dfrac{-7 \pm \sqrt{17}}{4}$ (4) $-4, \ \dfrac{1}{2}$

35 In the accompanying diagram of quadrilateral $QRST$, $\overline{RS} \cong \overline{ST}$, $\overline{SR} \perp \overline{QR}$, and $\overline{ST} \perp \overline{QT}$.

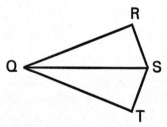

Which method of proof may be used to prove $\triangle QRS \cong \triangle QTS$?

(1) HL   (3) AAS

(2) SAS   (4) ASA

Part II

Answer *three* questions from this part. Clearly indicate the necessary steps, including appropriate formula substitutions, diagrams, graphs, charts, etc. Calculations that may be obtained by mental arithmetic or the calculator do not need to be shown. [30]

36 Answer *a*, *b*, and *c* for all values of *x* for which these expressions are defined.

a Find the value of $\dfrac{(x+1)^2}{x^2-1}$ if $x = 1.02$. [2]

b Find the positive value of *x* to the *nearest thousandth*:

$$\frac{1}{x} = \frac{x+1}{1}$$ [5]

c Solve for all values of *x* in simplest radical form:

$$\frac{x+2}{4} = \frac{2}{x-2}$$ [3]

37 Triangle *ABC* has coordinates $A(1,0)$, $B(7,4)$, and $C(5,7)$.

a On graph paper, draw and label $\triangle ABC$. [1]

b Graph and state the coordinates of $\triangle A'B'C'$, the image of $\triangle ABC$ after a reflection in the origin. [3]

c Graph and state the coordinates of $\triangle A''B''C''$, the image of $\triangle A'B'C'$ under the translation $(x,y) \rightarrow (x+1, y+5)$. [3]

d Write an equation of the line containing $\overline{A''B''}$. [3]

38 Solve the following system of equations algebraically and check:

$$y = x^2 - 6x + 5$$
$$y + 7 = 2x \qquad [8,2]$$

39 In the accompanying diagram of rhombus $ABCD$, $m\angle CAB = 25°$ and $AC = 18$.

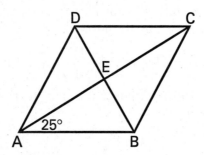

Find, to the *nearest tenth*, the

a perimeter of $ABCD$ [6]

b length of $\overline{BD}$ [4]

40 The vertices of $\triangle NYS$ are $N(-2,-1)$, $Y(0,10)$, and $S(10,5)$. The coordinates of point T are $(4,2)$.

a Prove that $\overline{YT}$ is a median. [2]

b Prove that $\overline{YT}$ is an altitude. [4]

c Find the area of $\triangle NYS$. [4]

Part III

Answer *one* question from this part. Clearly indicate the necessary steps, including appropriate formula substitutions, diagrams, graphs, charts, etc. Calculations that may be obtained by mental arithmetic or the calculator do not need to be shown. [10]

41 Given: $\angle 1 \cong \angle 2$ and $\overline{DB} \perp \overline{AC}$.

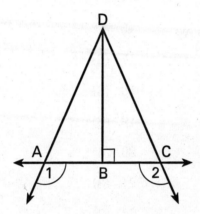

Prove: $\triangle ABD \cong \triangle CBD$ [10]

42 Given: $\sim G \rightarrow F$
$\sim(E \wedge F)$
$\sim E \rightarrow \sim D$
A
$(B \wedge C) \rightarrow D$
$A \rightarrow (B \wedge C)$

Prove: G [10]

ANSWER KEY

Part I

1. 11	13. 20	25. (1)
2. 53	14. $\frac{x}{2}$	26. (3)
3. 6	15. 5	27. (2)
4. 15	16. 4	28. (1)
5. 35	17. (2)	29. (2)
6. 1	18. (3)	30. (1)
7. 8	19. (3)	31. (4)
8. 12	20. (4)	32. (4)
9. $\overline{AC}$	21. (3)	33. (4)
10. 110	22. (2)	34. (1)
11. –3	23. (4)	35. (1)
12. 3	24. (2)	

EXPLANATIONS:
JUNE 1996

Part I

1. 11

Don't let the symbols freak you out. This function question defines what the little sun symbol means, so all you have to do is plug in $a = 5$ and $b = 7$:

$$a \odot b = a^2 - 2b$$
$$5 \odot 7 = 5^2 - 2(7) = 25 - 14 = 11$$

2. 53

Use the "inverse tangent" button on your calculator. (It usually says "tan⁻¹" and involves the "second function" button.) Type 1.34 into your calculator and press "tan⁻¹." You should get 53.267. When you round this off to the nearest degree, as instructed, you get 53°.

3. 6

The identity element of a system is the one that never changes any of the characters in the original row. Look at the diagram below:

★	2	4	6	8
2	4	8	2	6
4	8	6	4	2
6	2	4	6	8
8	6	2	8	4

The row of numbers at the top of the chart is identical to the row that lines up with the number 6. (The same is true about the column headed by the 6.) Therefore, 6 is the identity element.

4. 15

The first thing to realize is that $\triangle ABC$ and $\triangle ADE$ are similar. This is true because $\overline{DE}$ is parallel to $\overline{BC}$; there are two pairs of corresponding angles: $\angle ADE$ and $\angle ABC$, and $\angle AED$ and $\angle ACB$:

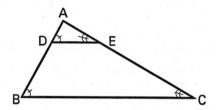

The two triangles are similar because of the Angle-Angle Rule of Similarity.

Corresponding sides of the two triangles are proportional. Set up the following proportion:

$$\frac{AD}{AB} = \frac{AE}{AC}$$

$$\frac{10}{24} = \frac{AE}{36}$$

Cross-multiply, and you're done:

$$24 \times (AE) = 36 \times 10$$
$$24(AE) = 360$$
$$AE = 15$$

5. 35

You should recognize this term as a calculation involving the Combinations Formula, which looks like this:

To find the value of $_7C_3$, plug in $n = 7$ and $r = 3$:

$$_7C_3 = \frac{7!}{3!\,4!} = \frac{7 \times 6 \times 5 \times 4 \times 3 \times 2 \times 1}{3 \times 2 \times 1 \times (4 \times 3 \times 2 \times 1)}$$

$$= \frac{7 \times 6 \times 5}{3 \times 2 \times 1} = \frac{210}{6} = 35$$

6. 1

Since 2 is a root of the equation, you can plug in 2 for x to determine the value of k:

$$x^2 + kx = 6$$
$$(2)^2 + k(2) = 6$$
$$4 + 2k = 6$$
$$2k = 2$$
$$k = 1$$

7. 8

Whenever two fractions are equal to each other, you can cross-multiply:

$$\frac{16}{y} = \frac{y}{4}$$
$$y(y) = 16 \times 4$$
$$y^2 = 64$$
$$y = \{8, -8\}$$

Since they only want the positive root, toss out –8. To check your work, plug 8 back into the equation and make sure it works.

8. 12

The formula to follow is a variation of the permutations rule. To find the number of possible arrangements of the letters in a word with n letters, in which one letter appears p times (remember that p is greater than 1), the formula looks like this:

$$\frac{n!}{p!}$$

NINE has 4 letters, but there are two N's. Therefore, you can express the number of arrangements as:

$$\frac{4!}{2!} = \frac{4 \times 3 \times 2 \times 1}{2 \times 1} = 4 \times 3 = 12$$

9. $\overline{AC}$

You can combine the two inequalities in the question like this:

$$m\angle B > m\angle C > m\angle A$$

From this, you can tell that $\angle B$ is the biggest angle in the triangle, which might look something like this:

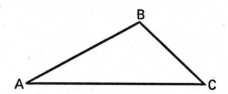

Since $\angle B$ is the biggest angle, the side opposite $\angle B$, or side $\overline{AC}$, is the biggest side.

10. 110

It's important to note two things here: (1) the diagonals of a rhombus bisect the angles from which they're drawn, and (2) opposite sides of a rhombus are parallel. Since $\overline{AC}$ is a diagonal, it bisects $\angle DAB$. Therefore, $m\angle DAC = 35$.

Since $\overline{DC}$ is parallel to $\overline{AB}$, $\angle DCA$ and $\angle CAB$ are alternate interior angles—which always have the same measure. Therefore, $m\angle DCA = 35$.

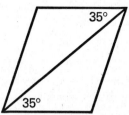

Now use the Rule of 180 to solve for the measure of $\angle ADC$.

$$m\angle ADC + m\angle DCA + m\angle CAD = 180$$
$$m\angle ADC + 35 + 35 = 180$$
$$m\angle ADC = 110$$

11. –3

Put the equation into $y = mx + b$ format, and you'll be able to figure out the slope of the line (the m) right away.

$$3x + y = 4$$
$$y = -3x + 4$$

The slope of this line is –3 (and the y-intercept, in case you were wondering, is 4).

12. 3

If you don't recognize the equation of a circle in the question, the phrase "locus of points at a given distance . . . from the origin" should have given you a hint. The distance, d, represents the radius of the circle. Since the equation of a circle centered at the origin is

$x^2 + y^2 = r^2$ (in which r is the radius), $r^2 = 9$. Although r could equal 3 or –3, you're looking for a distance, which can't be negative. Therefore, $d = 3$.

13. 20

Plot your points on the coordinate axes like this:

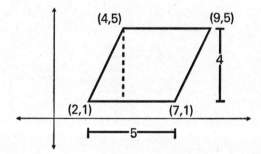

The formula for the area of a parallelogram is bh, in which b is the base and h is the perpendicular height. The base equals the distance between (2,1) and (7,1), or 5 units. (Don't bother with the distance formula here; just count the boxes on the graph paper.) The height of the parallelogram is 4 units (see diagram). The area equals 5×4, or 20 square units.

14. $\dfrac{x}{2}$

You can't do anything until the two fractions have the same denominator, which is the lowest common denominator (LCD) of 6 and 3. The LCD is 6, so you don't have to do anything to the first fraction. To make the second fraction compatible, multiply both the top and bottom by 2:

$$\frac{x(2)}{3(2)} = \frac{2x}{6}$$

Now you can subtract the fractions and reduce:

$$\frac{5x}{6} - \frac{2x}{6} = \frac{3x}{6}$$

$$\frac{3x}{6} = \frac{x}{2}$$

15. **5**

This question gives you all but one component of the slope formula, so use it to figure out the value of y. The slope formula is:

$$m = \frac{y_2 - y_1}{x_2 - x_1}$$

Let $(x_1, y_1) = (1, 3)$ and $(x_2, y_2) = (2, y)$. The slope, m, is 2:

$$2 = \frac{y - 3}{2 - 1}$$

$$2 = \frac{y - 3}{1}$$

$$2 = y - 3$$

$$5 = y$$

Check your work by plugging 5 back into the original equation.

16. **4**

The diagonal of a square bisects the square into two 45:45:90 tri angles. Since the ratio of the sides of a 45:45:90 triangle is $1:1:\sqrt{2}$, the legs of the triangle (which are also the sides of the square) must be 4 units long.

If you never studied 45:45:90 triangles (or you forgot the ratio), there's another way to solve this:

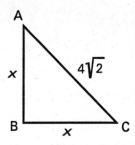

Diagonal $\overline{AC}$ is the hypotenuse of $\triangle ABC$. Since the sides of a square have equal length, $AB = BC$. Label each of these x, then use the Pythagorean Theorem:

$$(AB)^2 + (BC)^2 = (AC)^2$$
$$x^2 + x^2 = (4\sqrt{2})^2$$
$$2x^2 = 32$$
$$x^2 = 16$$
$$x = \{4, -4\}$$

Since all distances are positive, toss out −4.

Multiple Choice

17. (2)

First, factor out an x:

$$x^3 - 9x = x(x^2 - 9)$$

This doesn't match any of the answer choices, so keep going. The term $x^2 - 9$ is a difference of squares:

$$(x^2 - 9) = (x + 3)(x - 3)$$

The complete factorization looks like this:

$$x(x^2 - 9) = x(x + 3)(x - 3)$$

Note: You're in Process Of Elimination country now. One of the best ways to arrive at the right answer is to prove the others wrong.

Combine the terms in each of the other answer choices and see what you get:

    (1) $x(x-3) = x^2 - 3x$

    (3) $(x+3)(x-3) = (x^2-9)$

    (4) $x(x+3) = x^2 + 3x$

None of these matches the original term $x^3 - 9x$, so cross 'em off.

18. (3)

The formula for a circle with center (h, k) and radius r is:

$$(x-h)^2 + (y-k)^2 = r^2$$

The key to the rest of this question is remembering the minus signs in the formula above. If you substitute $(-2, k)$ for (h, k) in the formula, you get:

$$[x - (-2)]^2 + (y-k)^2 = 25$$
$$(x+2)^2 + (y-k)^2 = 25$$

When you compare this equation to the one in the problem, k must equal 3.

19. (3)

Once you determine the value of y, the rest of the problem falls into place. Since $\angle BCA$ and $\angle ACD$ are supplementary, their sum must be 180:

$$6y + 3y = 180$$
$$9y = 180$$
$$y = 20$$

Now you know that m $\angle BCA = 6 \times 20$, or $120°$, and m $\angle B = 2 \times 20$, or $40°$.

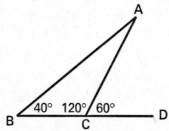

Use the Rule of 180:
$$m\angle A + m\angle B + m\angle BCA = 180$$
$$m\angle A + 40 + 120 = 180$$
$$m\angle A = 20$$

20. (4)

The median of a triangle connects a vertex to the midpoint of the side opposite that vertex. The median from vertex A, therefore, connects point A to the midpoint of $\overline{BC}$.

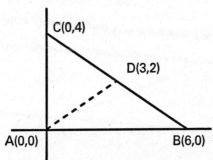

You can find the midpoint by using the midpoint formula:

$$(\overline{x}, \overline{y}) = \left(\frac{x_1 + x_2}{2}, \frac{y_1 + y_2}{2} \right)$$

Let $(x_1, y_1) = B(6, 0)$ and $(x_2, y_2) = C(0, 4)$:

$$(\overline{x}, \overline{y}) = \left(\frac{6+0}{2}, \frac{0+4}{2} \right) = \left(\frac{6}{2}, \frac{4}{2} \right) = (3, 2)$$

21. (3)

Whenever you see the negation of a logic statement in parentheses, think of De Morgan's Laws.

$$\sim(a \wedge b) \rightarrow \sim a \vee \sim b$$

This basically means that when you negate a parenthetical statement with a "$\wedge$" or "$\vee$" in it, negate each symbol and turn the symbol upside down:

$$\sim(\sim m \wedge n) \rightarrow \sim(\sim m) \vee \sim n$$

Since $\sim(\sim m)$ is the same thing as m (because of the rule of double negation), you can rewrite the statement as: $m \vee \sim n$.

22. (2)

In order to find the image of a point (x, y) after a translation that maps it onto its image $(x - 2, y + 3)$, subtract 2 from the x-coordinate and add 3 to the y-coordinate. Under this translation, the point $(7,2)$ would be mapped onto point $(7 - 2, 2 + 3)$, or $(5,5)$.

23. (4)

This figure represents three similar right triangles (FUR, URN, and FUN), and all their corresponding sides are proportional to each other. For this problem, consider $\triangle FUR$ and $\triangle URN$ and set up a proportion. In $\triangle FUR$, UR is the long leg and FR is the short leg; in $\triangle URN$, RN is the long leg and UR is the short leg.

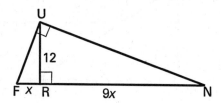

You can't enter numbers right away. You're looking for FR, so label it x. The ratio of FR to RN is 1:9, so the length of RN is $9x$. Now you can set up the proportion:

$$\frac{FR}{UR} = \frac{UR}{RN}$$

$$\frac{x}{12} = \frac{12}{9x}$$

When you cross-multiply, you'll get:

$$9x(x) = 12 \times 12$$
$$9x^2 = 144$$
$$x^2 = 16$$
$$x = \{4, -4\}$$

Since you're looking for a distance, which can't be a negative value, eliminate -4.

24. (2)

When two lines are perpendicular, their slopes are negative reciprocals. That is, their product is –1. Therefore, if line l has a slope of $\frac{3}{5}$, the slope of line m is the negative reciprocal of $\frac{3}{5}$, or $-\frac{5}{3}$.

25. (1)

Consider the angles of $\triangle ECD$ to figure out this one. Start with $\angle ECD$, which is supplementary to $\angle ECA$. Since m $\angle ECA$ = 65, it must be true that m $\angle ECD$ = 115. Further, $\angle FEB$ and $\angle CED$ are vertical angles, so they have the same measure, 25°.

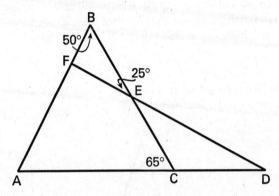

Use the Rule of 180 to determine m $\angle D$:

$$\text{m}\angle ECD + \text{m}\angle CED + \text{m}\angle D = 180$$
$$115 + 25 + \text{m}\angle D = 180$$
$$\text{m}\angle D = 40$$

26. (3)

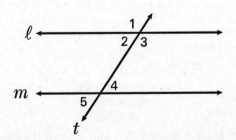

In this diagram, $\angle 1$ and $\angle 2$ are supplementary, so m $\angle 1$ + m $\angle 2$ = 180. Similarly, $\angle 2$ and $\angle 3$ are supplementary, so m $\angle 2$ + m $\angle 3$ = 180. Using the transitive property of addition, you can prove that m $\angle 1$ + m $\angle 2$ = m $\angle 2$ + m $\angle 3$; both quantities equal 180, so they must be equal to each other.

Process Of Elimination can help you eliminate the other choices rather easily.

27. (2)

Answer choice (2) has the problem. You're given the expression $a \vee b$, which means "either a is true OR b is true." The next statement you're given is $\sim b$, which means that b is not true. From these two statements, you have to conclude that a is true, not false.

28. (1)

Be careful as you set up this diagram. The way it's worded (and the vocabulary) can be confusing:

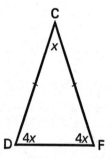

You want the measure of the vertex angle (which is $\angle C$, at the top of the triangle), so set it equal to x. The base angles of an isosceles triangle are equal to each other, and each of them is four times the measure of the vertex angle; set each of them equal to $4x$. Now use the Rule of 180:

$$\text{m} \angle C + \text{m} \angle D + \text{m} \angle F = 180$$
$$x + 4x + 4x = 180$$
$$9x = 180$$
$$x = 20$$

The vertex angle measures 20°, and each of the base angles measures 4×20, or 80°.

29. (2)

After a reflection in the line $y = x$, the x- and y-coordinates are interchanged. In other words, $r_{y\,=\,x}\,(x, y) \to (y, x)$. The image of point $R(-4, 3)$ is $R'(3, -4)$.

30. (1)

The locus of points that are equidistant from two points A and B is another name for the perpendicular bisector of $\overline{AB}$. The line $y = 4$ is horizontal and passes through the point $(0,4)$. If you plot each pair of points on the same coordinate axes, you'll see that the line is equidistant from points $(0,0)$ and $(0,8)$:

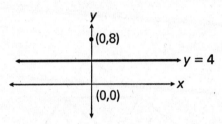

31. (4)

Make your diagram look something like this, and be sure to make $\overline{BD}$ *smaller* than $\overline{DC}$:

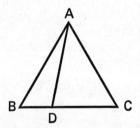

You can't be sure of the exact length of $\overline{AD}$, but any line segment drawn within $\triangle ABC$ has to be shorter than a side of the triangle. Therefore, it must be true that AC is greater than AD.

32. (4)

When a parabola appears in the form $y = ax^2 + bx + c$, the equation of its axis of symmetry is:

$$x = -\frac{b}{2a}$$

For this parabola, $a = 2$, $b = 7$, and $c = -5$. Thus the axis of symmetry is:

$$x = -\frac{-7}{2(2)} = -\frac{7}{4}$$

The equation for the line is $x = -\frac{7}{4}$.

33. (4)

It doesn't matter if the triangle is scalene or not. The answer will always be the same:

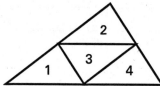

When you connect the midpoints of the sides of *any* triangle, you'll cut the triangle into four smaller congruent triangles.

34. (1)

No matter how hard you try, you can't factor the equation. (Cross off answer choices (3) and (4), which assume you can.) You have to use the Quadratic Formula:

$$x = \frac{-b \pm \sqrt{b^2 - 4ac}}{2a}$$

In the equation $2x^2 - 7x + 4 = 0$, $a = 2$, $b = -7$, and $c = 4$:

$$x = \frac{-(-7) \pm \sqrt{(-7)^2 - 4(2)(4)}}{2(2)} = \frac{7 \pm \sqrt{49 - 32}}{4} = \frac{7 \pm \sqrt{17}}{4}$$

35. (1)

The hint for this one is the information that $\overline{SR}\perp\overline{QR}$ and $\overline{ST}\perp\overline{QT}$:

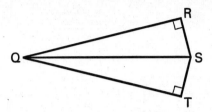

The fact that $\triangle QRS$ and $\triangle QTS$ are right triangles should get you thinking about Hypotenuse-Leg. $\overline{RS}$ and $\overline{ST}$ are congruent corresponding legs; $\overline{QS}$ is the hypotenuse of each triangle; and $QS = QS$ because of the Reflexive Property of Equality.

You can prove that the two triangles are congruent by using Hypotenuse-Leg.

Part II

36. *a* 101

Plug in 1.02 for x, and keep your calculator handy:

$$\frac{(1.02+1)^2}{(1.02)^2-1} = \frac{(2.02)^2}{(1.02)^2-1} = \frac{4.0804}{1.0404-1} = \frac{4.0804}{0.0404} = 101$$

Note: You can also make this problem a lot simpler by reducing

$$\frac{(x+1)^2}{x^2-1} \quad \text{to} \quad \frac{x+1}{x-1}$$

***b* 0.618**

Whenever two fractions are equal to each other, you can cross-multiply:

$$\frac{1}{x} = \frac{x+1}{1}$$
$$x(x+1) = 1 \times 1$$
$$x^2 + x = 1$$
$$x^2 + x - 1 = 0$$

You can't factor the equation (the fact that your answer has to be rounded to nearest hundredth should be a clue), so you have to use the Quadratic Formula:

$$x = \frac{-b \pm \sqrt{b^2 - 4ac}}{2a}$$

In the equation $x^2 + x - 1 = 0$, $a = 1$, $b = 1$, and $c = -1$:

$$x = \frac{-1 \pm \sqrt{(1)^2 - 4(1)(-1)}}{2(1)} = \frac{-1 \pm \sqrt{1 + 4}}{2} = \frac{-1 + \sqrt{5}}{2}, \frac{-1 - \sqrt{5}}{2}$$

Use your calculator to determine that $\sqrt{5} = 2.2361$, and plug that into the two terms:

$$x = \frac{-1 + 2.2361}{2}$$

$$= \frac{1.2361}{2}$$

$$= 0.61805$$

$$x = \frac{-1 - 2.2361}{2}$$

$$= \frac{-3.2361}{2}$$

$$= -1.61805$$

The positive root is 0.61801, which rounds to 0.618.

c $x = \left\{ 2\sqrt{3}, -2\sqrt{3} \right\}$

Here are two more fractions to cross-multiply:

$$\frac{x + 2}{4} = \frac{2}{x - 2}$$
$$(x + 2)(x - 2) = 4 \times 2$$
$$x^2 - 4 = 8$$
$$x^2 = 12$$
$$x = \sqrt{12}, -\sqrt{12}$$

To reduce the radical, factor out a perfect square:

$$\sqrt{12} = \sqrt{4 \times 3} = \sqrt{4} \times \sqrt{3} = 2\sqrt{3}$$

They want *all* possible values of x, so don't forget to include the negative value.

37. a

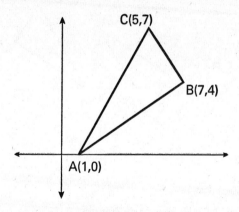

b $A'(-1,0)$, $B'(-7,-4)$, $C'(5,-7)$

After a reflection in the origin, the x- and y-coordinates are both negated. In other words, $r_{(0,0)}(x, y) \rightarrow (-x, -y)$. Your diagram should look like this:

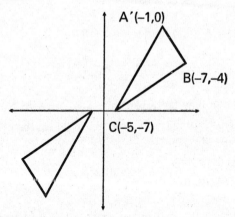

c $A''(0,5)$, $B''(-6,1)$, $C''(-4,-2)$

To find the images of the three points under the translation $(x + 1, y + 5)$, add 1 to each of the x-coordinates and add 5 to each of the y-coordinates. Now, your diagram should look like this:

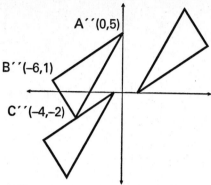

A''(0,5)
B''(-6,1)
C''(-4,-2)

d $y = \dfrac{2}{3}x + 5$

Your goal is to find the slope (represented by m) and the y-intercept (the b) and plug them into the form $y = mx + b$. Find the slope of $\overline{A''B''}$ using the slope formula:

$$m = \frac{y_2 - y_1}{x_2 - x_1}$$

Let $(x_1, y_1) = A''(0, 5)$ and $(x_2, y_2) = B''(-6, 1)$:

$$m = \frac{1-5}{-6-0} = \frac{-4}{-6} = \frac{2}{3}$$

Your equation now looks like this: $y = \dfrac{2}{3}x + b$. Now you have to find b by substituting the coordinates of one of the points. It doesn't matter which one you use, but $A''(0,5)$ works out easier:

$$5 = \frac{2}{3}(0) + b$$
$$5 = b$$

Note: If you realize that $(0,5)$ is the y-intercept of the line, you can recognize that $b = 5$.

The equation now reads: $y = \dfrac{2}{3}x + 5$.

38. (2,–3) and (6,5)

<u>Algebraic Method</u>

If you choose to do this algebraically, put the second equation into $y = mx + b$ form by subtracting 7 from both sides:

$$y + 7 = 2x$$
$$y = 2x - 7$$

Now you have two separate equations that are equal to y. Therefore, you can set them equal to each other:

$$x^2 - 6x + 5 = 2x - 7$$
$$x^2 - 8x + 5 = -7$$
$$x^2 \quad 8x + 12 = 0$$

Factor the binomial and set each of the factors equal to zero:

$$(x - 6)(x - 2) = 0$$
$$x = \{6, 2\}$$

If $x = 6$, then $y = (6)^2 - 6(6) + 5$, or 5. One point of intersection is (6,5).

If $x = 2$, then $y = (2)^2 - 6(2) + 5$, or –3. The other point of intersection is (2,–3).

<u>Graphing Method</u>

If you choose to graph the two equations, you may want to find the axis of symmetry of the parabola before you make your T-chart. The formula for the axis of symmetry is:

$$x = -\frac{b}{2a}$$

In the parabola $y = x^2 - 6x + 5$, $a = 1$, $b = -6$, and $c = 5$:

$$x = -\frac{-6}{2(1)} = \frac{6}{2} = 3$$

The axis of symmetry is $x = 3$, so choose three x-coordinates to the left of 3 (0, 1, and 2) and three to the right of 3 (4, 5, and 6). Plug each one in for x in the equation to find the y-coordinate for each point. For example, if $x = 0$, then $y = (0)^2 - 6(0) + 5$, or 5. The coordinate pair is (0,5). The remaining points are below. Use some of the same x-coordinates when you graph the line $y = 2x - 7$ on the same set of axes. The graphs look like this:

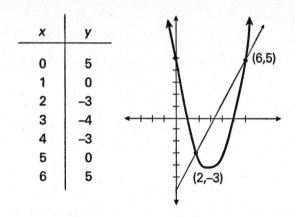

x	y
0	5
1	0
2	-3
3	-4
4	-3
5	0
6	5

The points of intersection are marked.

Note: Be sure to check the points by plugging them back into the system. Otherwise you'll lose 2 points.

39. *a*   39.7

Before you get started, it's important to note that the diagonals of a rhombus (a) bisect each other, and (*b*) are perpendicular.

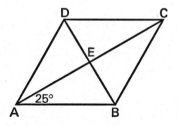

Therefore, $\triangle AED$ is a right triangle, and AE equals half the length of diagonal $\overline{AC}$ (or 9). To find the perimeter of the rhombus, you can find the length of one side and multiply it by 4 (because a rhombus has four equal sides). You know the length of the leg adjacent to $\angle EAB$, and you want to find the length of the hypotenuse $\overline{AB}$. Therefore, it's time to trig; use cosine (the **CAH** in **SOHCAHTOA**):

$$\text{cosine} = \frac{\text{adjacent}}{\text{hypotenuse}}$$

$$\cos \angle EAB = \frac{EA}{AB}$$

$$\cos 25° = \frac{9}{AB}$$

Since the cosine of 25° is 0.9063, enter that into the equation and cross-multiply:

$$0.9063 = \frac{9}{AB}$$
$$AB(0.9063) = 9$$
$$AB = \frac{9}{0.9063}$$
$$AB = 9.93$$

Multiply this value by 4 and you get 39.72, which becomes 39.7 when you round it off to the nearest tenth.

b   8.4

You could use trig here, but you don't have to. Instead, focus on right triangle ABE; you can find EB using the Pythagorean Theorem:

$$(AE)^2 + (EB)^2 = (AB)^2$$
$$9^2 + (EB)^2 = (9.93)^2$$
$$81 + (EB)^2 = 98.6$$
$$(EB)^2 = 17.6$$
$$EB = 4.19$$

Multiply this by 2 (because $\overline{BD}$ is twice as long as $\overline{EB}$), and you get 8.38. This rounds to 8.4.

40. *a*

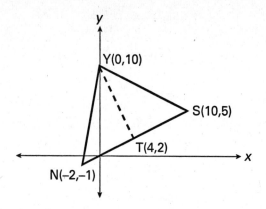

A median of a triangle is a line segment drawn from one vertex to the midpoint of the side opposite that vertex. To prove that $\overline{YT}$ is a median, you have to show that $T(4, 2)$ is the midpoint of $\overline{NS}$ using the midpoint formula:

$$(\overline{x}, \overline{y}) = \left(\frac{x_1 + x_2}{2}, \frac{y_1 + y_2}{2}\right)$$

Let $(x_1, y_1) = N(-2, -1)$ and $(x_2, y_2) = S(10,5)$:

$$(\overline{x}, \overline{y}) = \left(\frac{-2 + 10}{2}, \frac{-1 + 5}{2}\right) = \left(\frac{8}{2}, \frac{4}{2}\right) = (4, 2)$$

b

An altitude of a triangle is a line segment drawn from one vertex perpendicular to the side opposite that vertex. To prove that $\overline{YT}$ is an altitude, you have to find the slopes of $\overline{YT}$ and $\overline{NS}$ and show that the two slopes are negative reciprocals (that is, the product of their slopes equals -1). Use the slope formula:

$$m = \frac{y_2 - y_1}{x_2 - x_1}$$

To find the slope of $\overline{YT}$, let $(x_1,y_1) = Y(0,10)$ and $(x_2,y_2) = T(4,2)$. To find the slope of $\overline{NS}$, let $(x_1,y_1) = N(-2,-1)$ and $(x_2, y_2) = S(10,5)$:

Slope of $\overline{YT}$:

$$m = \frac{2-10}{4-0}$$

$$= \frac{-8}{4} = -2$$

Slope of $\overline{NS}$:

$$m = \frac{5-(-1)}{10-(-2)}$$

$$= \frac{6}{12} = \frac{1}{2}$$

Since $-2 \times \frac{1}{2} = -1$, the two segments are perpendicular, and $\overline{YT}$ is an altitude.

c   60

The best way to determine the area of $\triangle NYS$ is to work indirectly. First, circumscribe a rectangle on the triangle in the diagram, like so:

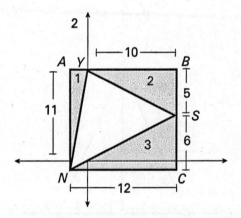

Triangle NYS is now surrounded by three other triangles within the rectangle. If you subtract their combined area from the area of the rectangle, the result will be the area of $\triangle NYS$.

The area of rectangle $ABCN$ is 12×11, or 132 square units. Now, calculate the area of the three outer triangles, using the formula $A = \frac{1}{2}bh$:

Triangle 1: $A = \frac{1}{2}(11)(2) = 11$

Triangle 2: $A = \frac{1}{2}(10)(5) = 25$

Triangle 3: $A = \frac{1}{2}(6)(12) = 36$

Total area of three triangles: $11 + 25 + 36 = 72$.

Subtract 72 from the rectangle's area (132), and you're left with 60 square units.

Part III

41.

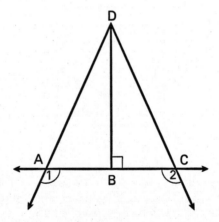

Prove: $\triangle ABD \cong \triangle CBD$

The plan: They've given you information about two angles, and the two triangles share a side. Use AAS.

Statements	Reasons
1. m∠1 = m∠2	1. Given
2. ∠1 is supplementary to ∠DAB; ∠2 is supplementary to ∠DCB.	2. Definition of supplementary angles
3. ∠DAB ≅ ∠DCB	3. Angles that are supplementary to congruent angles are congruent.
4. $\overline{DB} \perp \overline{AC}$	4. Given
5. ∠ABD and ∠CBD are right angles.	5. Definition of perpendicular lines
6. ∠ABD ≅ ∠CBD	6. All right angles are congruent.
7. $\overline{DB} \cong \overline{DB}$	7. Reflexive Property of Congruence
8. △ABD ≅ △CBD	8. AAS ≅ AAS

42.

Here's the proof. Notice that the letters in the statements proceed alphabetically (more or less).

Statements	Reasons
1. $A \rightarrow (B \wedge C)$; $(B \wedge C) \rightarrow D$	1. Given
2. $A \rightarrow D$	2. Chain Rule (1)
3. A	3. Given
4. D	4. Law of Detachment (2, 3)
5. $\sim E \rightarrow \sim D$	5. Given
6. E	6. Law of *Modus Tollens* (4, 5)
7. $\sim(E \wedge F)$	7. Given
8. $\sim E \vee \sim F$	8. De Morgan's Laws
9. $\sim F$	9. Law of Disjunctive Inference (6, 8)
10. $\sim G \rightarrow F$	10. Given
11. G	11. Law of *Modus Tollens* (9, 10)

EXAMINATION:
AUGUST 1996

Part I

Answer 30 questions from this part. Each correct answer will receive 2 credits. No partial credit will be allowed. Write your answers in the spaces provided on the separate answer sheet. Where applicable, answers may be left in terms of π or in radical form. [60]

1 The table below defines the operation × for the set $F = \{1, -1, y, -y\}$. What is the value of $(-1) \times y$?

×	1	−1	y	$-y$
1	1	8	y	$-y$
−1	−1	1	$-y$	y
y	y	$-y$	−1	1
$-y$	$-y$	y	1	−1

2 In the accompanying diagram, $\overline{OA} \perp \overline{OB}$ and $\overline{OD} \perp \overline{OC}$. If m$\angle 3 = 39$, what is m$\angle 1$?

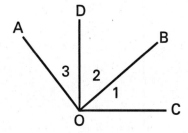

3 Solve for x: $\dfrac{x-2}{2} = \dfrac{x-1}{3}$

4 In rectangle $ABCD$, diagonal $AC = 20$ and segment $\overline{EF}$ joins the midpoints of $\overline{AB}$ and $\overline{BC}$, respectively. Find the length of $\overline{EF}$.

5 What is the total number of points equidistant from two intersecting lines and two centimeters from the point of intersection?

6 In $\triangle ABC$, m $\angle A = 40$ and the measure of an exterior angle B is $120°$. Which side is the longest in $\triangle ABC$?

7 What is the negative root of the equation $x^2 - x - 2 = 0$?

8 In the accompanying diagram of $\triangle ABC$, $\overline{DE} \| \overline{BC}$, $AD = 3$, $AB = 9$, and $AE = 5$. Find EC.

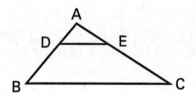

9 What is the image of $(-2,5)$ after a reflection in the x-axis?

10 In rhombus $ABCD$, $AB = 2$ and $BC = x + 8$. find the length of $\overline{BC}$.

11 Express $\dfrac{x + 2}{3} + \dfrac{x - 3}{4}$ as a single fraction in simplest form.

12 In the accompanying diagram, $\overleftrightarrow{AB}$ is parallel to $\overleftrightarrow{CD}$, AED is a transversal, and $\overline{CE}$ is drawn. If m$\angle CED = 60$, m$\angle DAB = 2x$, and m$\angle DCE = 3x$, find x.

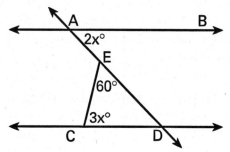

13 Find the area of a triangle whose vertices are $(-2,0)$, $(-2,6)$, and $(5,0)$.

14 If the endpoints of the diameter of a circle are $(3,1)$ and $(6,5)$, find the length of the diameter.

15 The coordinates of the midpoint of line segment $\overline{AB}$ are $(-2,4)$. If the coordinates of point A are $(7,10)$, find the coordinates of point B.

16 The sides of a triangle measure 5, 9, and 10. Find the perimeter of a similar triangle whose longest side measures 15.

Directions (17-34): For *each* question chosen, write the *numeral* preceding the word or expression that best completes the statement or answers the question.

17 The coordinates of point (x,y) after a reflection in the origin can be represented by

(1) (x,y) (3) $(x,-y)$

(2) $(-x,y)$ (4) $(-x,-y)$

18 If the length of the hypotenuse of a right triangle is 4 and the length of one leg is 2, what is the length of the other leg?

(1) 12   (3) $\sqrt{12}$

(2) 20   (4) $\sqrt{20}$

19 Which equation represents the line whose slope is –2 and that passes through point $(0,3)$?

(1) $y = -2x + 3$ (3) $y = 3x - 2$

(2) $y = -2x - 3$ (4) $y = 2x + 3$

20 If the lengths of two sides of a triangle are 4 and 8, the length of the third side may *not* be

(1) 5   (3) 7

(2) 6   (4) 4

21 What is the length of an altitude of an equilateral triangle whose side measures 6?

(1) $3\sqrt{2}$ (3) 3

(2) $3\sqrt{3}$ (4) $6\sqrt{3}$

22 What is the slope of a line parallel to the line whose equation is $y = 5x + 4$?

(1) $-\dfrac{4}{5}$ (3) 5

(2) $-\dfrac{5}{4}$ (4) 4

23 What is the contrapositive of $c \to (d \vee e)$?

(1) $\sim c \to \sim(d \vee e)$ (3) $\sim(d \vee e) \to \sim c$

(2) $c \to \sim(d \vee e)$ (4) $(d \vee e) \to c$

24 What is an equation of the circle whose center is $(-3,1)$ and whose radius is 10?

(1) $(x + 3)^2 + (y - 1)^2 = 10$

(2) $(x + 3)^2 + (y - 1)^2 = 100$

(3) $(x - 3)^2 + (y + 1)^2 = 10$

(4) $(x - 3)^2 + (y + 1)^2 = 100$

25 Given three premises: $A \to \sim C$, $\sim C \to R$, and $\sim R$. Which conclusion *must* be true?

(1) R (3) $A \wedge C$

(2) $\sim C$ (4) $\sim A$

26 In $\triangle ABC$, m $\angle A = 41$ and m $\angle B = 48$. What kind of triangle is $\triangle ABC$?

(1) right   (3) isosceles

(2) obtuse   (4) acute

27 If the altitude is drawn to the hypotenuse of a right triangle, then the two triangles formed are *always*

(1) congruent   (3) isosceles

(2) equal   (4) similar

28 The number of sides of a regular polygon whose interior angles each measure 108° is
(1) 5 (3) 7
(2) 6 (4) 4

29 Two triangles have altitudes of equal length. If the areas of these triangles have the ratio 3:4, then the bases of these triangles have the ratio
(1) 3:4 (3) $\sqrt{3}$:2
(2) 9:16 (4) $\frac{3}{2}$:2

30 In isosceles triangle ABC, $AC = BC = 20$, $m\angle A = 68$, and $\overline{CD}$ is the altitude to side $\overline{AB}$. What is the length of $\overline{CD}$ to the *nearest tenth*?
(1) 49.5 (3) 10.6
(2) 18.5 (4) 7.5

31 If quadrilateral $ABCD$ is a parallelogram, which statement must be true?

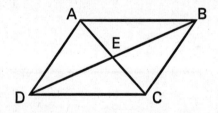

(1) $\overline{AC} \perp \overline{BD}$
(2) $\overline{AC} \cong \overline{BD}$
(3) $\overline{AC}$ bisects $\angle DAB$ and $\angle BCD$.
(4) $\overline{AC}$ and $\overline{BD}$ bisect each other.

32 How many different seven-letter arrangements can be formed from the letters in the word "GENESIS"?

(1) 210   (3) 1260
(2) 840   (4) 5040

33 What is the turning point of the graph of the function $y = x^2 - 6x + 2$?

(1) (3,−7)   (3) (3,11)
(2) (−3,−7)   (4) (−3,11)

34 A biology class has eight students. How many different lab groups may be formed that will consist of three students?

(1) (3,−7)   (3) (3,11)
(2) (−3,−7)   (4) (−3,11)

Directions (35): Leave all constructions lines on the answer sheet.

35 *On the answer sheet*, costruct a line through point P that is perpendicular to $\overline{AB}$.

Part II

Answer *three* questions from this part. Clearly indicate the necessary steps, including appropriate formula substitutions, diagrams, graphs, charts, etc. Calculations that may be obtained by mental arithmetic or the calculator do not need to be shown. [30]

36 *a* On graph paper, draw the graph of the equation $y = 2x^2 - 4x - 3$ for all values of x in the interval $-2 \leq x \leq 4$. [6]

    b On the same set of axes, draw the reflection of the graph of the equation $y = 2x^2 - 4x - 3$ in the y-axis. [2]

    c What is the equation of the axis of symmetry of the graph in part *b*? [2]

37 Solve the following system of equations algebraically and check.

$$x^2 + y^2 + 4x$$
$$x + y = 0 \qquad [8,2]$$

38 Given: $\qquad M \rightarrow N$

$\qquad\qquad\qquad \sim M \rightarrow P$

$\qquad\qquad\qquad (L \wedge N) \rightarrow R$

$\qquad\qquad\qquad \sim R$

$\qquad\qquad\qquad L$

Using the laws of inference, prove P. [10]

39 Answer both *a* and *b* for all values of *x* for which these expressions are defined.

a Solve for *x* to the nearest hundredth:

$$\frac{1}{x-1} = \frac{x+4}{5} \quad [6]$$

b Simplify:

$$\frac{x^2-4}{x^2+4x+4} \cdot \frac{x^2+2x}{x^2} \quad [4]$$

40 In the accompanying diagram of rectangle *ABCD*, diagonal $\overline{AC}$ is drawn, DE = 8, $\overline{DE} \perp \overline{AC}$ and m∠DAC = 55. Find the area of rectangle *ABCD* to the *nearest integer*. [10]

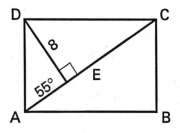

Part III

Answer *one* question from this part. Clearly indicate the necessary steps, including appropriate formula substitutions, diagrams, graphs, charts, etc. Calculations that may be obtained by mental arithmetic or the calculator do not need to be shown.   [10]

41 Given parallelogram DEBK, $\overline{BC} \cong \overline{DA}$ and $\overline{DJ} \cong \overline{BL}$.

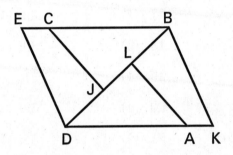

Prove: $\overline{CJ} \cong \overline{AL}$ [10]

42 Quadrilateral *ABCD* has coordinates *A*(0,–6), *B*(5,–1), *C*(3,3), and *D*(–1,1).

Using coordinate geometry, prove that

a at least two consecutive sides are not congruent.   [5]

b the diagonals, $\overline{AC}$ and $\overline{BD}$, are perpendicular.   [5]

ANSWER KEY

Part I

1. $-y$
2. 39
3. 4
4. 10
5. 4
6. $\overline{AB}$
7. -1
8. 10
9. $(-2,-5)$
10. 18
11. $\dfrac{7x-1}{12}$
12. 24

13. 21
14. 5
15. $(-11,-2)$
16. 36
17. (4)
18. (3)
19. (1)
20. (4)
21. (2)
22. (3)
23. (2)
24. (2)

25. (4)
26. (2)
27. (4)
28. (1)
29. (1)
30. (2)
31. (4)
32. (3)
33. (1)
34. (1)
35. Construction

EXPLANATIONS: AUGUST 1996

Part I

1. $-y$

Find "–1" in the far left column, and run your finger along that row until you get to the column headed by "$-y$":

x	1	–1	y	$-y$
1	1	–1	y	$-y$
–1	–1	1	y	y
y	y	$-y$	–1	1
$-y$	$-y$	y	1	–1

At this point of intersection, you'll find "$-y$". Therefore, the value of $(-1) \times y$ is $-y$.

2. 39

Since $\overline{OA} \perp \overline{OB}$, m$\angle AOB$ = 90 and $\angle 3$ is complementary to $\angle 2$. Using the same line of reasoning, you also know that m$\angle DOC$ = 90 because $\overline{OD} \perp \overline{OC}$. Therefore, $\angle 1$ is complementary to $\angle 2$. If two angles are complementary to the same angle, they are congruent. Therefore, m$\angle 1$ = m$\angle 3$ = 39°.

3. 4

Whenever two fractions are equal to each other, you can cross-multiply:

$$\frac{x-2}{2} = \frac{x-1}{3}$$
$$3(x - 2) = 2(x - 1)$$
$$3x - 6 = 2x - 2$$
$$3x = 2x + 4$$
$$x = 4$$

To check your math, plug 4 back into the equation and make sure it works.

4. 10

Draw your diagram first:

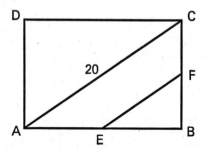

Focus your attention on $\triangle ABC$. A segment that connects the midpoints of two sides of a triangle is half the length of the third side. E is the midpoint of $\overline{AB}$ and F is the midpoint of $\overline{BC}$, so EF must be half of AC. Since $AC = 20$, $EF = 10$.

5. 4

This problem can appear rather difficult at first glance. It requires a lot of visualization. Look at the intersecting lines first. These lines form two sets of vertical angles, and the locus of points equidistant from the two lines are the lines that bisect those vertical angles:

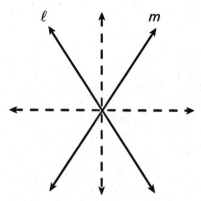

Now look at the point of intersection. The locus of points that are two centimeters from that point is a circle with a 2-inch radius:

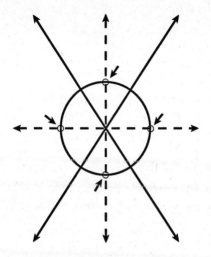

There are four points of intersection

6. $\overline{AB}$

The triangle looks like this:

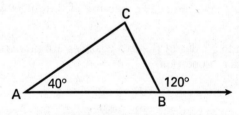

Since ∠CBD is an exterior angle, it's supplementary to ∠ABC. Therefore, m∠ABC = 60. Use the Rule of 180 to determine the measure of ∠C:

$$m\angle A + m\angle ABC + m\angle C = 180$$
$$40 + 60 + m\angle C = 180$$
$$m\angle C = 80$$

Since ∠C is the largest angle in the triangle, the side opposite ∠C, or $\overline{AB}$, is the largest side.

7. –1

To find the roots of this binomial, you have to factor it and set each factor equal to zero:

$$x^2 - x - 2 = 0$$
$$(x - 2)(x + 1) = 0$$
$$x = \{2, -1\}$$

They want the *negative* root, so get rid of the 2. To check your math, plug –1 back into the equation and make sure it works.

8. 10

The first thing to realize is that $\triangle ABC$ and $\triangle ADE$ are similar. This is true because $\overline{DE}$ is parallel to $\overline{BC}$; there are two pairs of corresponding angles: $\angle ADE$ and $\angle ABC$, and $\angle AED$ and $\angle ACB$:

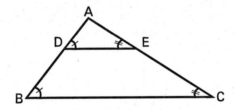

The two triangles are similar because of the Angle-Angle Rule of Similarity.

Corresponding sides of the two triangles are proportional. Set up the following proportion:

$$\frac{AD}{AB} = \frac{AE}{AC}$$
$$\frac{3}{9} = \frac{5}{AC}$$

Now cross-multiply:

$$3 \times (AC) = 9 \times 5$$
$$3(AC) = 45$$
$$AC = 15$$

Hold it. There's still a bit more work to do. You know that $AE = 5$ and $AC = 15$. Since $AC = AE + EC$, the length of $\overline{EC}$ is 10.

9.(–2, –5)

After a reflection in the x-axis, the x-coordinate remains the same and the y-coordinate is negated. In other words, $r_{x\text{-axis}} (x, y) \rightarrow (x, -y)$. The image of point (–2, 5) is (–2, –5). To check your work, you can plot both points:

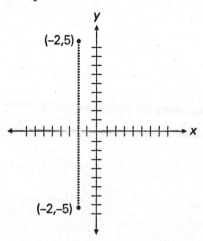

10. 18

A rhombus is a quadrilateral with four equal sides; therefore, $AB = BC$. Set the two equal to each other and solve for x:

$$2x - 2 = x + 8$$
$$2x = x + 10$$
$$x = 10$$

To find the length of $\overline{BC}$, substitute 10 for x and solve:

$$BC = x + 8 = 10 + 8 = 18$$

11. $\dfrac{7x - 1}{12}$

You can't do anything until the two fractions have the same denominator. This number is the lowest common denominator (LCD) of 4 and 3, which is 12. To make two fractions compatible, multiply the top and bottom of the first fraction by 4:

$$\frac{4}{4} \; \frac{(x - 3)}{4} = \frac{4x + 8}{12}$$

Multiply the top and bottom of the second fraction by 3.

$$\frac{3}{3} \cdot \frac{(x-3)}{4} = \frac{3x-9}{12}$$

Now you can add the fractions:

$$\frac{4x+8}{12} + \frac{3x-9}{12} = \frac{4x+8+3x-9}{12} = \frac{7x-1}{12}$$

12. 24

Ignore segment $\overline{EC}$ for a moment. Since $\overline{AB}$ is parallel to $\overline{CD}$, and is a transversal, $\angle BAE$ and $\angle EDC$ are alternate interior angles (which have the same measure). Therefore, m $\angle BAE$ = m $\angle EDC$ = 2x.

Now look at $\triangle EDC$. You can find the value of x by using the Rule of 180:

$$\text{m}\angle CED + \text{m}\angle EDC + \text{m}\angle DCE = 180.$$
$$60 + 2x + 3x = 180$$
$$60 + 5x = 180$$
$$5x = 120$$
$$x = 24$$

13. 21

Plot the three points like this:

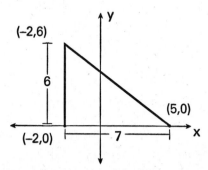

This triangle is a right triangle with a base of length 7 and a height of length 6. Use the formula for the area of a triangle (in which b represents the length of the base and h represents the height):

$$A = \frac{1}{2}bh = \frac{1}{2}(7)(6) = \frac{1}{2}(42) = 21$$

14. 5

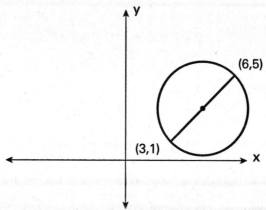

To find the distance between the two points, use the distance formula:

$$d = \sqrt{(x_2 - x_1)^2 + (y_2 - y_1)^2}$$

Let $(x_1, y_1) = (3,1)$ and $(x_2, y_2) = (6,5)$:

$$d = \sqrt{(6 - 3)^2 + (5 - 1)^2} = \sqrt{3^2 + 4^2} = \sqrt{9 + 16} = \sqrt{25} = 5$$

15. (−11, −2)

The formula for the midpoint of a line segment is:

$$(\bar{x}, \bar{y}) = \left(\frac{x_1 + x_2}{2}, \frac{y_1 + y_2}{2} \right)$$

You'll have to use this formula a little differently by solving for each coordinate individually. Let $(\bar{x}, \bar{y}) = (-2,4), (x_1, y_1) = A(7,10)$ and $(x_2, y_2) = B(x,y)$:

$$\bar{x} = \frac{x_1 + x_2}{2}$$

$$-2 = \frac{7 + x}{2}$$

$$-4 = 7 + x$$

$$-11 = x$$

$$\bar{y} = \frac{y_1 + y_2}{2}$$

$$4 = \frac{10 + y}{2}$$

$$8 = 10 + y$$

$$-2 = y$$

The coordinates of point B are $(-11, -2)$.

16. 36

Any time a problem involves similar triangles, all you do is set up a proportion. The key is lining up the corresponding sides. First draw a diagram:

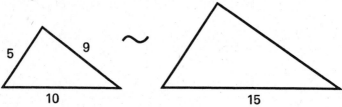

When two triangles are similar, the ratio of the lengths of any pair of corresponding sides is the same as the ratio of the perimeters. The longest side of the smaller triangle is 10 units long, and its counterpart of the larger triangle is 15 units long. The perimeter of the small triangle is 5 + 9 + 10, or 24, and you want to find the perimeter of the larger triangle (call it P). Your proportion should look like this:

$$\frac{10}{15} = \frac{24}{P}$$

Cross-multiply and solve:

$$10P = 360$$
$$x = 36$$

Multiple Choice

17. (4)

After a reflection in the origin, both the x- and y-coordinates are negated. In other words, $r_{(0.0)}(x, y) \rightarrow (-x, -y)$.

Process Of Elimination is a great help as well. Answer choice (1) is out right away, because the only point that remains unchanged after a reflection in the origin is the origin itself. Answer choice (2) represents a reflection in the y-axis, and answer choice (3) represents a reflection in the x-axis.

18. (3)

Draw your right triangle like this:

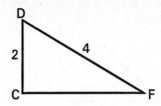

Use the Pythagorean Theorem to find the length of the other leg:

$$(DC)^2 + (CF)^2 = (DF)^2$$
$$2^2 + (CF)^2 = 4^2$$
$$4 + (CF)^2 = 16$$
$$(CF)^2 = 12$$
$$CF = \sqrt{12}$$

Note: If you know square roots well, you can do this problem without doing any math at all. The other leg $\overline{(CF)}$ has to be shorter than the hypotenuse, which is only 4 units long. You can cross off answer choices (1) and (2) right off the bat; since $\sqrt{20}$ is also greater than 4, you can eliminate answer choice (4) as well.

19. (1)

When a line is written in the $y = mx + b$ format (as all the answer choices are), the m represents the slope of the line and the b represents the y-intercept. Since the slope of the line in question is -2, get rid of answer choices (3) and (4). Also, (0,3) is the y-intercept of the line because the x-coordinate is 0; b must equal 3. The correct equation is $y = -2x + 3$.

20. (4)

Given the lengths of two sides of a triangle, the length of the third side has to be smaller than the sum of the other two sides and larger than their difference:

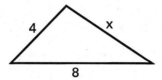

In this case, the length of the third side must be:

$$(8 - 4) < x < (8 + 4)$$
$$4 < x < 12$$

Answer choice (4) is not within this range. Remember that x must be *greater than* 4. It can't be equal to 4.

21. (2)

Draw an equilateral triangle and an altitude from the top vertex, like this:

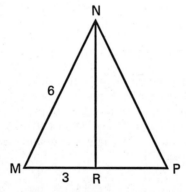

Look at $\triangle MNR$. An altitude of an equilateral triangle bisects the vertex andle and is perpendicular to the opposite side. Therefore, $\triangle MNR$ is a 30:60:90 triangle. In a 30:60:90 triangle, the short leg is half as long as the hypotenuse. Therefore, $MR = 3$. The length of the long leg of a 30:60:90 triangle equals the length of the short leg times $\sqrt{3}$. The length of $\overline{NR}$ (the altitude of the triangle) is $3\sqrt{3}$.

Note: If you never learned the special relationship between the sides of a 30:60:90 triangle, you can always use the Pythagorean Theorem. The key is realizing that $MR = 3$.

22. (2)

When a line is writtenin the $y = mx + b$ forma, the m represents the slope of the line and the b represents the y-intercept. Therefore, the slope of the line $y = 5x + 4$ is 5. All parallel lines have the same slope, so any line parallel to this line will also have a slope of 5.

23. (3)

The contrapositive is an abbreviated reference to the Law of Contrapositive Inference (what we like to call the Flip-and-Negate Rule). To find the contrapositive of and "if-then" statement, flip the two terms to the opposite side of the arrow and negate them both, (For example, $a \rightarrow b$ becomes $b \rightarrow \sim a$.) In this case, the contrapositive of $c \rightarrow (d \vee e)$ is :

$$\sim(d \vee e) \rightarrow \sim c$$

24. (2)

Use the formula for a circle, and remember that (h, k) is the center and r is the radius:

$$(x - h)^2 + (y - k)^2 = r^2$$
$$[x - (-3)]^2 + (y - 1)^2 = 10^2$$
$$(x + 3)^2 + (y - 1)^2 = 100$$

Since the formula involves r^2 and not r, you should recognize that the formula will equal 100, not 10. Therefore, eliminate answer choices (1) and (3).

25. (4)

Using the Chain Rule, you can combine the first two statements: if $A \rightarrow \sim C$ and $\sim C \rightarrow R$, you can conclude that $A \rightarrow R$. The contrapositive of this statement is $\sim R \rightarrow \sim A$ (if you're not sure why, see Question 23). The statement $\sim R$ is given, so it must be true that $\sim A$.

26. (2)

Use the Rule of 180 to determine the measure of ∠C:

$$m \angle A + m \angle B + m \angle C = 180$$
$$41 + 48 + m \angle C = 180$$
$$m \angle C = 91$$

The measure of ∠C is greater than 90, so it's an obtuse angle. Therefore, the triangle is also obtuse.

27. (4)

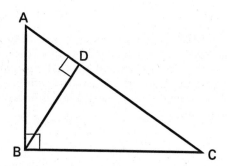

You've probably encountered this type of diagram before. When you draw the altitude of a right triangle, you cut the triangle into two smaller right triangles. All three of the triangles (the original one, △ABC, and the two smaller ones, △BDC and △BAD) are similar; all their corresponding sides and angles are proportional to each other.

Process of Elimination is also very helpful here. The two smaller triangles aren't the same size (so they can't be congruent); eliminate answer choices (1) and (2). None of the triangles is isosceles either.

28. (1)

Here's the formula for the number of degrees x in each angle of a regular polygon (n represents the number of sides in that polygon):

$$x = \frac{180(n-2)}{n}$$

Rather than plug in 108 for x, you might find it easier to plug each of the answer choices in for n. (**Note:** You don't have to bother with answer choice (4), because a regular polygon with four sides is a square, and you know each of those angles measures 90°.) Try plugging in $n = 5$:

$$x = \frac{180(5-2)}{5} = \frac{180(3)}{5} = \frac{540}{5} = 108$$

29. (1)

Create a diagram for each triangle and plug in some numbers. Let each altitude equal 5. The areas are in a ratio of 3:4, so let the area of triangle A equal 30 and the area of triangle B equal 40:

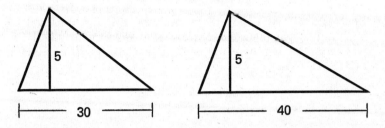

Using the formula for the area of a triangle $\left(A = \frac{1}{2}bh\right)$, find the length of each base:

$$A = \frac{1}{2}bh \qquad\qquad A = \frac{1}{2}bh$$

$$30 = \frac{1}{2}b(5) \qquad\qquad 40 = \frac{1}{2}b(5)$$

$$60 = 5b \qquad\qquad 80 = 5b$$

$$12 = b \qquad\qquad 16 = b$$

The two bases are 12 and 16, and the ratio 12:16 reduces to 3:4. Therefore, the two bases are in a 3:4 ratio.

30. (2)

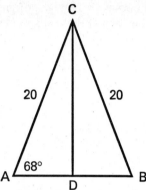

Look at $\triangle ACD$; you know the measure of $\angle A$. You also know the length of its hypotenuse, and you want to know the length of $\overline{CD}$, which is the leg opposite $\angle A$. Use the sine (the SOH in SOHCAHTOA):

$$\sin \angle A = \frac{\text{opposite}}{\text{hypotenuse}}$$

$$\sin 68° = \frac{CD}{20}$$

$$20(0.9272) = CD$$

$$18.544 = CD$$

When you round this off to the nearest *tenth*, your answer becomes 18.5.

31. (4)

The diagonals of a parallelogram bisect each other, so answer choice (4) is the correct response.

32. (3)

The formula to follow is a variation of the permutations rule. To find the number of possible arrangements of the letters in aËèrd with n letters, in which one letter appears p times and another letter appears q times (remember that p and q are greater than 1), the formula looks like this:

$$\frac{n!}{p!\,q!}$$

GENESIS has 7 letters, but there are two E's and two S's. Therefore, you can express the number of arrangements as:

$$\frac{7!}{2!\,2!} = \frac{7 \times 6 \times 5 \times 4 \times 3 \times 21}{2 \times 1 \times (2 \times 1)} = 7 \times 6 \times 5 \times 3 \times 2 = 1,260$$

33. (1)

To find the turning point, or vertex, of a parabola, find the x-coordinate of the turning point by using the formula for the axis of symmetry. When a parabola appears in the form $y + ax^2 + bx + c$ as this one does, the equation of its axis of symmetry is:

$$x = -\frac{b}{2a}$$

For this parabola, $a = 1$, $b = -6$, and $c = 2$. Thus the axis of symmetry is:

$$x = -\frac{-6}{2(1)} = \frac{6}{2} = 3$$

The x-coordinate of the vertex is 3, so you can eliminate answer choices (2) and (4). Now plug 3 into the equation of the parabola to determine the vertex's y-coordinate:

$$y = (3)^2 - 6(3) + 2 = 9 - 18 + 2 = -7$$

The coordinates of the turning point are $(3, -7)$.

34. (1)

This is a combinations problem (because the order of the students in the lab group doesn't matter), so use the combinations formula:

$$_nC_r = \frac{n!}{r!\,(n-r)!}$$

There are 8 students ($n = 8$), and you want to choose 3 of them ($r = 3$):

$$_8C_3 = \frac{8!}{3!\,5!} = \frac{8 \times 7 \times 6 \times 5 \times 43 \times 2 \times 1}{(3 \times 2 \times 1) \times (5 \times 4 \times 3 \times 2 \times 1)}$$

$$= \frac{8 \times 7 \times 6}{3 \times 2 \times 1} = \frac{336}{6} = 56$$

35.

With the point of your compass on P, draw an arc (#1) that intersects $\overline{AB}$ in two places:

Without altering the width of your compass, place its point on A and draw an arc (#2) on the side opposite P. Then put the point of your compass on B and draw another arc (#3) that intersects arc #2:

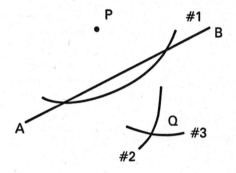

Draw the line defined by P and Q:

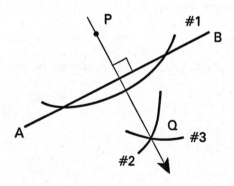

Part II

36. *a*

Start the graphing process by plugging the integers between –2 and 4, inclusive, into the equation for the parabola and determine its coordinates. For example, if $x = -2$, then $y = 2(-2)^2 - 4(-2) - 3$, or 13. Your first ordered pair is (–2,13). Here is the rest of your T-chart and the accompanying sketch:

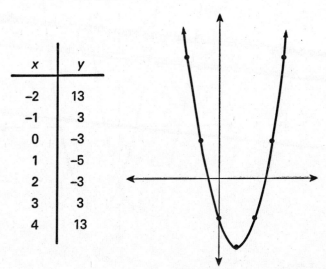

x	y
–2	13
–1	3
0	–3
1	–5
2	–3
3	3
4	13

b

After a reflection in the y-axis, the x-coordinate of each point is negated and the y-coordinate of each point remains unchanged. In other words, $r_{y\text{-axis}} (x,y) \rightarrow (-x,y)$. Use this formula to determine the new points; your graph should look like this:

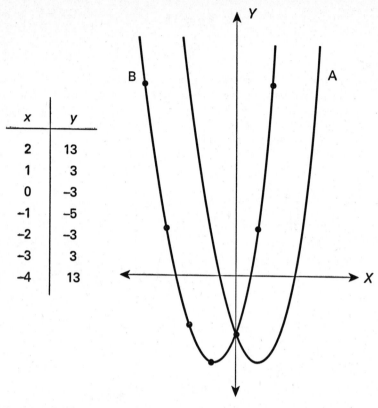

x	y
2	13
1	3
0	-3
-1	-5
-2	-3
-3	3
-4	13

c $x = -1$

When a parabola appears in the form $y = ax^2 + bx + c$ as this one does, the equation of its axis of symmetry is:

$$x = -\frac{b}{2a}$$

Don't bother trying to determine the equation of the parabola in Part B. You know the equation for the original parabola, so find its axis of symmetry and reflect it in the y-axis.

The original parabola appears in the standard $y = ax^2 + bx + c$ format; $a = 2$, $b = -4$, and $c = -3$. The axis of symmetry is:

$$x = -\frac{-4}{2(2)} = \frac{4}{4} = 1$$

Since the axis of symmetry of the original parabola is $x = 1$, the axis of symmetry of the parabola in Part B must be $x = -1$. (**Note:** You can always check the axis of symmetry by drawing it on the graph and making sure it cuts the second parabola in half.)

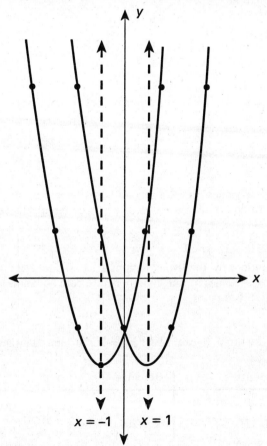

37. (0,0), (−2,2)

Look at the second equation first. If you subtract x from both sides, the equation $y + x = 0$ becomes $y = -x$. Substitute $-x$ for y in the first equation:

$$x^2 + y^2 + 4x = 0$$
$$x^2 + (-x)^2 + 4x = 0$$
$$x^2 + x^2 + 4x = 0$$
$$2x^2 + 4x = 0$$

Now you have to factor the equation, set each factor equal to 0, and solve for x:

$$2x(x + 2) = 0 \qquad\qquad x + 2 = 0$$
$$2x = 0 \qquad\qquad\qquad x = -2$$
$$x = 0$$

Plug these values into the second equation $y + x = 0$ (because it's easier) to determine the value of y in each circumstance:

$$y + 0 = 0 \qquad\qquad y + (-2) = 0$$
$$y = 0 \qquad\qquad\qquad y = 2$$

The solutions for the two equations are $(0,0)$ and $(-2,2)$. Be sure to check these values by plugging them into both equations. Otherwise, you'll lose two points.

38.

Here's the proof. (**Note:** These logical proofs normally appear in Part Three.)

Statements	Reasons
1. $(L \wedge N) \rightarrow R$; $\sim R$	1. Given
2. $\sim(L \wedge N)$	2. Law of *Modus Tollens*
3. $\sim L \vee \sim N$	3. De Morgan's Laws
4 L	4. Given
5. $\sim N$	5. Law of Disjunctive Inference (3, 4)
6. $M \rightarrow N$	6. Given
7. $\sim M$	7. Law of *Modus Tollens* (5, 6)
8. $\sim M \rightarrow P$	8. Given
9. P	9. Law of Detachment (7, 8)

(**Note:** If your class didn't cover the Law of *Modus Tollens*, you can use a combination of the Law of Contrapositive Inference and the Law of Detachment.)

39. *a* **1.85, –4.85**

Whenever two fractions are equal to each other, you can cross-multiply:

$$\frac{1}{x-1} = \frac{x+4}{5}$$
$$(x-1)(x+4) = 1 \times 5$$
$$x^2 + 3x - 9 = 0$$

Try as you might, you can't factor this equation. (That would be too easy!) The fact that you have to solve for x to the nearest hundredth should be a clue that you have to use the Quadratic Formula:

$$x = \frac{-b \pm \sqrt{b^2 - 4ac}}{2a}$$

In the equation, $a = 1$, $b = 3$, and $c = -9$:

$$x = \frac{-3 \pm \sqrt{(3)^2 - 4(1)(-9)}}{2(1)} = \frac{-3 \pm \sqrt{9 + 36}}{2} = \frac{-3 + \sqrt{45}}{2}, \frac{-3 - \sqrt{45}}{2}$$

Use your calculator to determine that $\sqrt{45} = 6.708$, and plug that into the two terms:

$$x = \frac{-3 + 6.708}{2} \qquad\qquad x = \frac{-3 - 6.708}{2}$$
$$= \frac{3.708}{2} \qquad\qquad\qquad = \frac{-9.708}{2}$$
$$= 1.854 \qquad\qquad\qquad\quad = -4.854$$

When you round each of these off to the nearest *hundredth*, your answers become 1.85 and –4.85. Check your math by plugging each of these values back into the original equation and making sure they work.

b $\dfrac{x-2}{x}$

Factor all the complex terms like this:

$$x^2 - 4 = (x + 2)(x - 2)$$
$$x^2 + 4x + 4 = (x + 2)(x + 2)$$
$$x^2 + 2x = x(x + 2)$$

Once you've factored these three terms, the problem looks like this:

$$\dfrac{(x + 2)(x - 2)}{(x + 2)(x + 2)} \cdot \dfrac{x(x + 2)}{x^2}$$

Cancel out all the factors that appear both on the top and on the bottom, and you're left with:

$$\dfrac{\cancel{(x + 2)}(x - 2)}{\cancel{(x + 2)}\cancel{(x + 2)}} \cdot \dfrac{\cancel{x}\cancel{(x + 2)}}{x^2} = \dfrac{x - 2}{x}$$

40. 136

To find the area of rectangle $ABCD$, you want to find AD and AB; once you find these lengths, you can use the formula for the area of a rectangle, $A = l \times w$.

To find AD, focus on right triangle EAD. You know the length of the leg opposite $\angle EAD$, and $\overline{AD}$ is the hypotenuse. It's time for more trig; use the sine (the SOH in SOHCAHTOA):

$$\sin \angle EAD = \dfrac{\text{opposite}}{\text{hypotenuse}}$$

$$\sin 55° = \dfrac{8}{AD}$$

$$0.8192 = \dfrac{8}{AD}$$

$$AD\,(0.8192) = 8$$

$$AD = \dfrac{8}{0.8192}$$

$$AD = 9.77$$

So far, so good. Now look at △*ABC*. Since *ABCD* is a rectangle, opposite sides $\overline{AD}$ and $\overline{BC}$ are parallel. Therefore, ∠*EAD* and ∠*BCA* are alternate interior angles, which have the same measure; m∠*BCA* = 55.

Furthermore, opposite sides of a rectangle are congruent, so *BC* also equals 9.77.

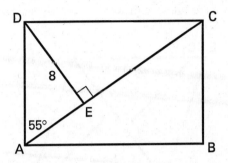

In △*ABC*, you know the length of the adjacent leg, and you want to find the length of the opposite leg, $\overline{AB}$ are parallel. Use the tangent (the TOA in SOHCAHTOA):

$$\tan \angle BCA = \frac{\text{opposite}}{\text{adjacent}}$$

$$\tan 55° = \frac{AB}{9.77}$$

$$1.4281 = \frac{AB}{9.77}$$

$$(1.4281)(9.77) = AB$$

$$AB = 13.95$$

Now you've found the rectangle's dimensions. When you multiply them together, you get:

$$13.95 \times 9.77 = 136.29$$

When you round this off to the nearest *integer*, your answer becomes

136.

Part III

41.

The plan: $\overline{CJ}$ and $\overline{AL}$ are corresponding sides of $\triangle JCB$ and $\triangle LAD$, respectively. They gave you some information to work with involving those two triangles, so prove that they're congruent using SAS, then use CPCTC.

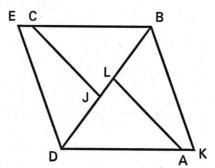

Statements	Reasons
1. $\overline{BC} \cong \overline{DA}$; DEBK is a parallelogram	1. Given
2. $\overline{EB}$ is parallel to $\overline{DK}$	2. Definition of a parallelogram
3. $\angle CBL \cong \angle JDA$	3. Alternate interior angles are congruent.
4. $\overline{DJ} \cong \overline{BL}$	4. Given
5. $DJ = BL$	5. Definition of congruence
6. $JL = JL$	6. Reflexive Property of Equality
7. $DJ + JL = BL + JL$	7. Additive Property of Equality
8. $DL = BJ$	8. Segment Addition Postulate
9. $\overline{DC} \cong \overline{BJ}$	9. Definition of congruence
10. $\triangle JCB \cong \triangle LAD$	10. SAS $\cong$ SAS
11. $\overline{CJ} \cong \overline{AL}$	11. Corresponding Parts of Congru-

42. *a*

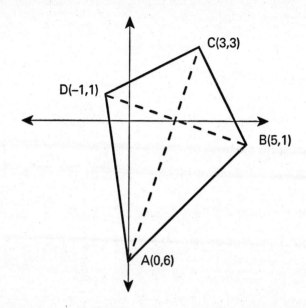

All you have to do is find one pair of consecutive sides that are not the same length. Use the distance formula to determine the lengths of the sides of $ABCD$; try $\overline{AB}$ and $\overline{BC}$ first:

$$d = \sqrt{(x_2 - x_1)^2 + (y_2 - y_1)^2}$$

$$
\begin{aligned}
AB &= \sqrt{(5 - 0)^2 + [-1 - (-6)]^2} \\
&= \sqrt{5^2 + 5^2} \\
&= \sqrt{25 + 25} \\
&= \sqrt{50} \\
&= 5\sqrt{2}
\end{aligned}
\qquad
\begin{aligned}
BC &= \sqrt{(3 - 5)^2 + [3 - (-1)]^2} \\
&= \sqrt{(-2)^2 + 4^2} \\
&= \sqrt{4 + 16} \\
&= \sqrt{20} \\
&= 2\sqrt{5}
\end{aligned}
$$

You've found a pair of consecutive sides that are not congruent.

b

If two lines are perpendicular, their slopes are negative reciprocals of each other. Determine the slope of each diagonal using the slope formula:

$$m = \frac{y_2 - y_1}{x_2 - x_1}$$

Slope of $\overline{AC}$:

$$m = \frac{3 - (-6)}{3 - 0}$$

$$= \frac{9}{3}$$

$$= 3$$

Slope of $\overline{BD}$:

$$m = \frac{1 - (-1)}{-1 - 5}$$

$$= \frac{2}{-6}$$

$$= -\frac{1}{3}$$

Since the slopes of the two diagonals are negative reciprocals $\left(3 \times -\frac{1}{3}\right) = -1$, the diagonals are perpendicular.

EXAMINATION:
JANUARY 1997

Part I

Answer 30 questions from this part. Each correct answer will receive 2 credits. No partial credit will be allowed. Write your answers in the spaces provided on the separate answer sheet. Where applicable, answers may be left in terms of π or in radical form. [60]

1 Using the accompanying table, solve for x if $x \circledast b = a$.

$\circledast$	a	b	c
a	a	b	c
b	b	a	c
c	c	c	b

2 In the accompanying table, $\triangle ABC$ is similar to $\triangle A'B'C'$, $AB = 14.4$, $BC = 8$, $CA = 12$, $A'B' = x$, and $B'C' = 4$. Find the value of x.

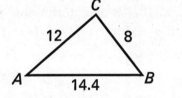 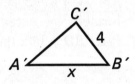

3 In the accompanying diagram, parallel lines $\overleftrightarrow{AB}$ and $\overleftrightarrow{CD}$ are intersected by $\overleftrightarrow{GH}$ at E and F, respectively. If m $\angle BEF = 5x - 10$ and m $\angle CFE = 4x + 20$, find x.

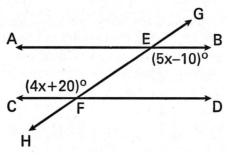

4 If tan $A = 0.5400$, find the measure of $\angle A$ to the *nearest degree*.

5 Find the length of a side of a square if two consecutive vertices have coordinates $(-2,6)$ and $(6,6)$.

6 In the accompanying diagram of isosceles triangle ABC, $CA = CB$ and $\angle CBD$ is an exterior angle formed by extending $\overline{AB}$ to point D. If m $\angle CBD = 130$, find m $\angle C$.

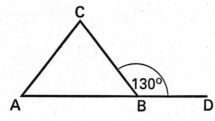

7 If $\overleftrightarrow{AB}$ intersects $\overleftrightarrow{CD}$ at E, m $\angle AEC = 3x$, and m $\angle AED = 5x - 60$, find the value of x.

8 Point (x,y) is the image of $(2,4)$ after a reflection in point $(5,6)$. In which quadrant does (x,y) lie?

9 In the accompanying diagram, $ABCD$ is a parallelogram, $\overline{EC} \perp \overline{DC}$, $\angle B \cong E$, and m $\angle A = 100$. Find m $\angle CDE$.

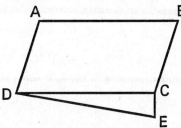

10 The lengths of the sides of $\triangle DEF$ are 6, 8, and 10. Find the perimeter of the triangle formed by connecting the midpoints of the sides of $\triangle DEF$.

11 The coordinates of the midpoint of line segment $\overline{AB}$ are $(1,2)$. If the coordinates of point A are $(1,0)$, find the coordinates of point B.

12 In $\triangle PQR$, $\angle Q \cong \angle R$. If $PQ = 10x - 14$, $PR = 2x + 50$, and $RQ = 4x - 30$, find the value of x.

13 What is the image of $(-2,4)$ after a reflection in the x-axis?

14 In rectangle $ABCD$, $\overline{AC}$ and $\overline{BD}$ are diagonals. If m $\angle 1$ = 55, find m $\angle ABD$.

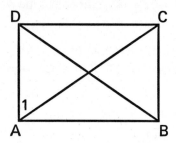

15 What is the slope of the line that passes through points $(-1,5)$ and $(2,3)$?

16 The coordinates of the turning point of the graph of the equation $y = x^2 - 2x - 8$ are $(1,k)$. What is the value of k?

Directions (17-35): For *each* question chosen, write the *numeral* preceding the word or expression that best completes the statement or answers the question.

17 Which equation represents the line that has a slope of $\frac{1}{2}$ and contains the point $(0,3)$?

(1) $y = \frac{1}{3}x + \frac{1}{2}$ (3) $y = \frac{3}{2}x$

(2) $y = 3x + \frac{1}{2}$ (4) $y = \frac{1}{2}x + 3$

18 If the measures of the angles in a triangle are in the ratio 3:4:5, the measure of an exterior angle of the triangle can *not* be

(1) 165° (3) 120°
(2) 135° (4) 105°

19 According to De Morgan's laws, which statement is logically equivalent to ~(p ∧ q)?

(1) ~(p ∨ ~q) (3) ~(p ∧ q)
(2) ~(p ∨ q) (4) ~(p ∧ ~q)

20 One angle of the triangle measures 30°. If the measures of the other two angles are in the ratio 3:7, the measure of the largest angle of the triangle is

(1) 15° (3) 126°
(2) 105° (4) 147°

21 In the accompanying diagram, *ABCD* is a rectangle, *E* is a point on $\overline{CD}$, m∠*DAE* = 30, and m∠*CBE* = 20.

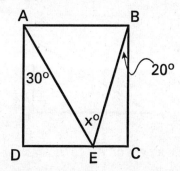

What is m∠*x*?

(1) 25 (3) 60
(2) 50 (4) 70

22 The graph of the equation $y = ax^2 + bx + c$, $a \neq 0$, forms

    (1) a circle   (3) a straight line
    (2) a parabola   (4) an ellipse

23 Which set of numbers can represent the lengths of the sides of a triangle?

    (1) {4,4,8}   (3) {3,5,7}
    (2) {3,9,14}   (4) {1,2,3}

24 Which is an equation of the line that passes through point (3,5) and is parallel to the x-axis?

    (1) $x = 3$ (3) $y = 5$
    (2) $x = 5$ (4) $y = 3$

25 What are the factors of $y^3 - 4y$?

    (1) $y(y - 2)(y - 2)$ (3) $y(y^2 + 1)(y - 4)$
    (2) $y(y + 4)(y - 4)$ (4) $y(y + 2)(y - 2)$

26 In the accompanying diagram of right triangle ABC, $AB = 4$ and $BC = 7$.

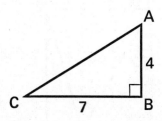

What is the length of $\overline{AC}$ to the *nearest hundredth*?

    (1) 5.74   (3) 8.06
    (2) 5.75   (4) 8.08

27 Which is the converse of the statement "If today is President's Day, then there is no school"?

(1) If there is school, then today is not President's Day.

(2) If there is no school, then today is President's Day.

(3) If today is President's Day, then there is school.

(4) If today is not President's Day, then there is school.

28 How many different eight-letter permutations can be formed from the letters in the word "PARALLEL"?

(1) $\dfrac{8!}{3!\,2!}$

(3) 360

(2) 8!

(4) $\dfrac{8!}{3!}$

29 Which equation describes the locus of points equidistant from $A(-3,2)$ and $B(-3,8)$?

(1) $x = -3$

(3) $x = 5$

(2) $y = -3$

(4) $y = 5$

30 A translation maps $A(1,2)$ onto $A'(-1,3)$. What are the coordinates of the image of the origin under the same translation?

(1) $(0,0)$

(3) $(-2,1)$

(2) $(2,-1)$

(4) $(-1,2)$

31 The solution set of the equation $x^2 + 5x = 0$ is

(1) $\{0\}$

(3) $\{-5\}$

(2) $\{5\}$

(4) $\{0,-5\}$

32 In the accompanying diagram of parallelogram *MATH*, m∠*T* = 100 and $\overline{SH}$ bisects ∠*MHT*.

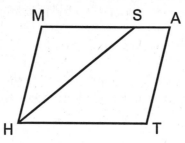

What is m∠*HSA*?

(1) 80   (3) 120

(2) 100   (4) 140

33 What are the roots of the equation $x^2 + 9x + 12 = 0$?

(1) $\dfrac{-9 \pm \sqrt{33}}{2}$ (3) $\dfrac{-9 \pm \sqrt{129}}{2}$

(2) $\dfrac{9 \pm \sqrt{33}}{2}$ (4) $\dfrac{9 \pm \sqrt{129}}{2}$

34 The vertices of trapezoid *ABCD* are *A*(–3,0), *B*(–3,4), *C*(2,4), and *D*(4,0). What is the area of trapezoid *ABCD*?

(1) 6   (3) 28

(2) 24   (4) 48

35　The accompanying diagram shows how $\triangle A'B'C'$ is constructed similar to $\triangle ABC$.

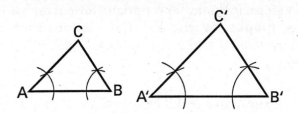

Which statement proves the construction?

(1) If two triangles are congruent, they are similar.

(2) If two triangles are similar, the angles of one triangle are congruent to the corresponding angles of the other triangle.

(3) Two triangles are similar if two angles of one triangle are congruent to two angles of the other triangle.

(4) The corresponding sides of two similar triangles are proportional.

Part II

Answer *three* questions from this part. Clearly indicate the necessary steps, including appropriate formula substitutions, diagrams, graphs, charts, etc. Calculations that may be obtained by mental arithmetic or the calculator do not need to be shown. [30]

36 Answer both *a* and *b* for all values of y for which these expressions are defined.

 a Express as a single fraction in lowest terms:

$$\frac{y-4}{2y} + \frac{3y-5}{5y} \quad [4]$$

 b Simplify:

$$\frac{y^2-7y+10}{5y-y^2} \div \frac{y^2-4}{25y^3} \quad [6]$$

37 In the accompanying diagram of isosceles triangle KLC, $\overline{LK} \cong \overline{LC}$, m$\angle K = 53$, altiutde $\overline{CA}$ is drawn to leg $\overline{LK}$, and $LA = 3$. Find the perimeter of $\triangle KLC$ to the *nearest integer*. [10]

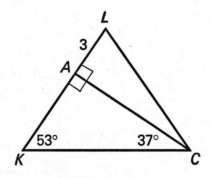

38 *a* On graph paper, draw the graph of the equation $y = -x^2 + 6x - 8$ for all values of x in the interval $0 \leq x \leq 6$. [6]

   b What is the maximum value of y in the equation $y = -x^2 + 6x - 8$? [2]

   c Write an equation of the line that passes through the turning point and is parallel to the x-axis. [2]

39 At a video rental store, Elyssa has only enough money to rent three videos. She has chosen four comedies, six dramas, and one mystery movie to consider.

   a How many different selections of three videos may she rent from the movies she has chosen? [2]

   b How many selections of three videos will consist of one comedy and two dramas? [3]

   c What is the probability that a selection of three videos will consist of one of each type of video? [3]

   d Elyssa decides to rent one comedy, one drama, and one mystery movie. In how many different orders may she view these videos? [2]

40 In the accompanying diagram of right triangle ABC, altitude is drawn to hypotenuse $\overline{BD}$, $AC = 20$, $AD < DC$, and $BD = 6$.

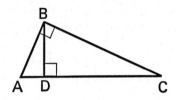

a If $AD = x$, express DC in terms of x. [1]

b Solve for x. [6]

c Find AB in simplest radical form. [3]

Part III

Answer *one* question from this part. Clearly indicate the necessary steps, including appropriate formula substitutions, diagrams, graphs, charts, etc. Calculations that may be obtained by mental arithmetic or the calculator do not need to be shown. [10]

41 Given: $\triangle ABC$; $\overline{BD}$ is both the median and the altitude to $\overline{AC}$.

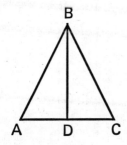

Prove: $\overline{BA} \cong \overline{BC}$ [10]

42 Quadrilateral $ABCD$ has vertices $A(-6,3)$, $B(-3,6)$, $C(9,6)$, and $D(-5,-8)$. Prove that quadrilateral $ABCD$ is

a a trapezoid [6]

b *not* an isosceles trapezoid [4]

ANSWER KEY

Part I

1. b

2. 7.2

3. 30

4. 28

5. 8

6. 80

7. 30

8. Quadrant I

9. 10

10. 12

11. (1,4)

12. 8

13. (−2,−4)

14. 35

15. $-\dfrac{2}{3}$

16. −9

17. (4)

18. (1)

19. (1)

20. (2)

21. (2)

22. (2)

23. (3)

24. (3)

25. (4)

26. (3)

27. (2)

28. (1)

29. (4)

30. (3)

31. (4)

32. (4)

33. (1)

34. (2)

35. construction

EXPLANATIONS: JANUARY 1997

Part I

1. *b*

Look along the top row for the column headed by *b*, and run your finger down that column until you find *a* in row *b*:

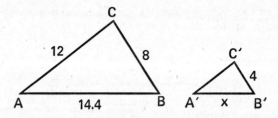

Since row *b* and column *b* intersect at point *a*, it must be true that $b \circledast b = a$. Therefore, $x = b$.

2. 7.2

Any time a problem involves similar triangles, all you do is set up a proportion. The key is lining up the corresponding sides:

This problem is a little easier because the sides are labeled. Set up a proportion involving corresponding sides:

$$\frac{AB}{A'B'} = \frac{BC}{B'C'}$$

$$\frac{14.4}{x} = \frac{8}{4}$$

Cross-multiply and solve:

$$8x = 57.6$$
$$x = 7.2$$

Note: An even quicker way to solve this is to realize that since $BC = 8$ and $B'C' = 4$, each side of $\triangle ABC$ is twice as long as its counterpart of $\triangle A'B'C'$. Therefore, $A'B'$ must be half as long as AB, or 7.2.

3. 30

Since $\overleftrightarrow{AB} \parallel \overleftrightarrow{CD}$, $\angle BEF$ and $\angle CFE$ are alternate interior angles, which must have the same measure. Set them equal to each other and solve for x:

$$m \angle BEF = m \angle CFE$$
$$5x - 10 = 4x + 20$$
$$5x = 4x + 30$$
$$x = 30$$

4. 28

Use the "inverse tangent" button on your calculator. (It usually says "tan⁻¹" and involves the "second function" button.) Once you're sure your calculator is in "degree mode," type in 0.54 and press "tan⁻¹." You should get 28.369. When you round this off to the nearest degree, as instructed, you get 28°.

5. 8

There are two ways to solve this one. The most direct way is to graph the two points like this:

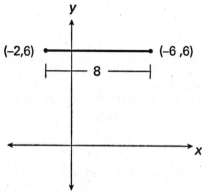

Since the segment between the points is horizontal, you can count the units between them; the segment is 8 units long.

Otherwise, you can use the distance formula.

6. 80

Since ∠*CBD* is an exterior angle, it's supplemental to ∠*ABC*. Therefore, m ∠*ABC* = 50°. It's also given that △*ABC* is isosceles. Since *CA* = *CB*, the angles opposite those sides, ∠*ABC* and ∠*A*, respectively, are also equal in measure. Therefore, m ∠*A* = 50°:

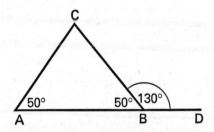

Now it's just a matter of using the Rule of 180:

$$m \angle A + m \angle ABC + m \angle C = 180$$
$$50 + 50 + m \angle C = 180$$
$$m \angle C = 80$$

7. 30

Always draw a diagram if you're not given one:

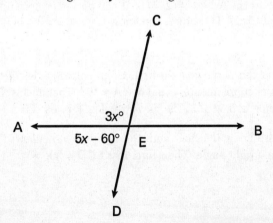

The diagram illustrates that $\angle AEC$ and $\angle AED$ are supplemental. Therefore, their sum is 180:

$$m \angle AEC + m \angle AED = 180$$
$$3x + (5x - 60) = 180$$
$$8x - 60 = 180$$
$$8x = 240$$
$$x = 30$$

8. Quadrant I

You might be tempted to find the exact coordinates of the image of (2, 4), but you don't have to. Both (2, 4) and (5, 6) are in the first quadrant:

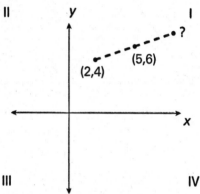

Whatever the image of (2, 4) is, it will be somewhere up and to the right of (5, 6). Therefore, you know it must also be in Quadrant I.

9. 10

Look at the parallelogram first. Consecutive angles of a parallelogram are supplementary, and $m \angle A = 100$. That means that $m \angle B = 80$. Since $\angle B \cong \angle E$, the measure of $\angle E$ is also 80.

You also know that $\overline{EC}$ and $\overline{DC}$ are perpendicular, so they intersect in a right angle. Therefore, $m \angle ECD = 90$.

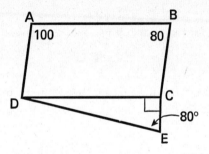

Now use the Rule of 180 in $\triangle CDE$:

$$m\angle CDE + m\angle ECD + m\angle E = 180$$
$$m\angle CDE + 90 + 80 = 180$$
$$m\angle CDE = 10$$

10. **12**

Draw a diagram, and let points X, Y, and Z be the midpoints of the three sides:

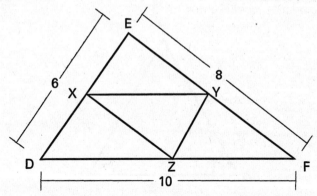

Look first at segment $\overline{XY}$. A segment that connects the midpoints of two sides of a triangle is half the length of the third side. Since X is the midpoint of $\overline{DE}$ and Y is the midpoint of $\overline{EF}$, $XY = 5$. Similarly, $XZ = 4$ and $YZ = 3$ Add these lengths up, and you'll have the perimeter of smaller triangle XYZ:

$$3 + 4 + 5 = 12$$

11. (1, 4)

The formula for the midpoint of a line segment is.

$$(\overline{x}, \overline{y}) = \left(\frac{x_1 + x_2}{2}, \frac{y_1 + y_2}{2} \right)$$

You'll have to use this formula a little differently by solving for each coordinate individually. Let $(\overline{x}, \overline{y}) = (1, 2)$, $(x_1, y_1) = A(1, 0)$ and $(x_2, y_2) = B(x, y)$:

$$\overline{x} = \frac{x_1 + x_2}{2}$$
$$1 = \frac{1 + x}{2}$$
$$2 = 1 + x$$
$$1 = x$$

$$\overline{y} = \frac{y_1 + y_2}{2}$$
$$2 = \frac{0 + y}{2}$$
$$4 = 0 + y$$
$$4 = y$$

The coordinates of point B are (1,4).

12. 8

The triangle looks like this:

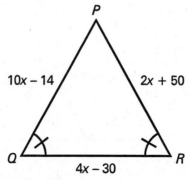

Since $\angle Q \cong \angle R$, the sides opposite those two angles ($\overline{PR}$ and $\overline{PQ}$, respectively) are equal in length. Therefore, you can set PR equal to PQ and solve for x:

$$PR = PQ$$
$$2x + 50 = 10x - 14$$
$$64 = 8x$$
$$8 = x$$

To check your math, plug 8 back into the original equation and make sure it works.

13. (–2, –4)

After a reflection in the x-axis, the x-coordinate remains the same and the y-coordinate is negated. In other words, $r_{x\text{-axis}} (x, y) \rightarrow (x, -y)$. The image of point (–2, 4) is (–2, –4)

14. 35

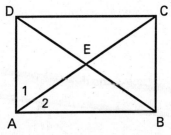

Since a rectangle has four right angles, m $\angle 1$ + m $\angle 2$ = 90. Therefore, m $\angle 2$ = 35. Now look at the point at which the diagonals intersect (labeled E). The diagonals of a rectangle are congruent, so AC = BD. The diagonals also bisect each other, so AE = EB. Now you know that $\triangle AEB$ is isosceles, which means that m $\angle 2$ = m $\angle ABD$. Thus, m $\angle ABD$ = 35.

15. $-\dfrac{2}{3}$

Determine the slope of the line using the slope formula:

$$m = \frac{y_2 - y_1}{x_2 - x_1}$$

Let $(x_1, y_1) = (-1, 5)$ and $(x_2, y_2) = (2, 3)$:

$$m = \frac{3 - 5}{2 - (-1)} = \frac{-2}{3} = -\frac{2}{3}$$

16. –9

The x-coordinate of the turning point is 1. To find the value of k, plug 1 in for x in the equation of the parabola:

$$y = (1)^2 - 2(1) - 8 = 1 - 2 - 8 = -9$$

Multiple Choice

17. (4)

Each of the answer choices is in the slope-intercept form $y = mx + b$, so m must be $\frac{1}{2}$ in the correct answer. This is true only in answer choice (4), so it must be correct. You don't even have to worry about (0, 3)!

18. (1)

First, find the measures of the three angles in the triangle using the Rule of 180:

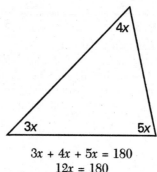

$$3x + 4x + 5x = 180$$
$$12x = 180$$
$$x = 15$$

The smallest angle measures 3×15, or 45. The other two angles must therefore measure 60 and 75. (The total of these three is 180—it's always good to check as you go.) An exterior angle must be supplemental to any of these angles. Therefore, the only possible measures of an exterior angle are 135, 120, and 105. The only answer not among these three is answer choice (1)

19. (1)

De Morgan's Laws state that when you negate a parenthetical statement with a " $\wedge$ " or " $\vee$ " in it, negate each symbol and turn the symbol upside down:

$$\sim (p \wedge q) \rightarrow \sim p \vee \sim q$$

Note: Since De Morgan's Laws always flip the symbol, you could have crossed off answer choices (3) and (4) right away

20. (2)

If one angle of a triangle measures 30°, the sum of the other two must be 150°. Since the ratio of the two unknown angles is 3:7, you can set up an equation like this:

$$3x + 7x = 150$$
$$10x = 150$$
$$x = 15$$

Hold it. Don't get careless and pick answer choice (1). The measure of the largest angle is 7×15, or 105°.

21. (2)

Since $ABCD$ is a rectangle, $\angle DAB$ and $\angle ABC$ are right angles; that is, m $\angle DAB$ = m $\angle ABC$ = 90. Therefore, m $\angle EAB$ = 60, and m $\angle ABE$ = 70:

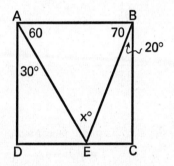

Now use the Rule of 180 within $\triangle ABE$:

$$\text{m} \angle EAB + \text{m} \angle ABE + \text{m} \angle BEA = 180$$
$$60 + 70 + x = 180$$
$$x = 50$$

22. (2)

The equation $y = ax^2 + bx + c$ is the standard form of a parabola. The reason they bother to include that $a \neq 0$ is that if $a = 0$, the x^2 term would drop out of the equation. The resulting equation, $y = bx + c$, is a line.

Note: POE works well here. For a graph to be a circle or an ellipse, both x and y must be squared. Eliminate answer choices (1) and (4). Further, the equation for a straight line has no squared terms, so you can get rid of answer choice (3).

23. (3)

Given the lengths of two sides of a triangle, the length of the third side has to be smaller than the sum of the other two sides. Since $4 + 4 = 8$ and $1 + 2 = 3$, answer choices (1) and (4) can't be correct. Answer choice (2) is also impossible, because $3 + 9$ is less than 14. The only choice left is answer choice (3).

24. (3)

Any line that is parallel to the x-axis is horizontal. The lines $x = 3$ and $x = 5$ are vertical, so you can eliminate answer choices (1) and (2). The y-coordinate of $(3, 5)$ is 5; the equation of the line that passes through the point $(3, 5)$ must be $y = 5$.

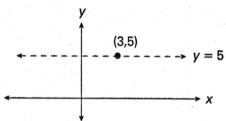

25. (4)

To break $y^3 - 4y$ down to its parts, factor a y out of each term first:
$$y^3 - 4y = y(y^2 - 4)$$
The term in parentheses is a difference of squares, so it breaks down like this:
$$y(y^2 - 4) = y(y - 2)(y + 2)$$

26. (3)

Use the Pythagorean Theorem:

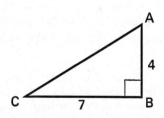

$$(AC)^2 = (AB)^2 + (CB)^2$$
$$(AC)^2 = 4^2 + 7^2$$
$$(AC)^2 = 16 + 49$$
$$(AC)^2 = 65$$
$$AC = \sqrt{65}$$

Use your calculator to find the square root of 65, which is 8.0622. Rounded to the nearest *hundredth*, your answer is 8.06.

Note: Use POE to eliminate answer choices (1) and (2); since the hypotenuse of a right triangle is always the largest of the three sides, AC must be greater than 7.

27. (2)

To find the converse of a conditional statement (otherwise known as an "if-then" statement), just switch the order of the statement. In other words, the converse of $p \rightarrow q$ is $q \rightarrow p$.

Therefore, the converse of "If today is Presidents' Day, then there is no school" is "If there is no school, then today is Presidents' Day."

28. (1)

The formula to follow is a variation of the permutations rule. To find the number of possible arrangements of the letters in a word with n letters, in which one letter appears p times and another letter appears q times (remember that p and q are greater than 1), the formula looks like this:

$$\frac{n!}{p!\, q!}$$

PARALLEL has 8 3 letters, but there are three L's and two A's. Therefore, you can express the number of arrangements as:

$$\frac{8!}{3!\, 2!}$$

29. (4)

To help you out, plot the two points like this:

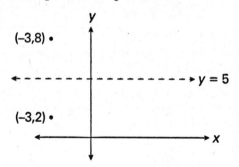

The locus of points between A and B is a horizontal line, so eliminate the vertical lines in answer choices (1) and (3). You have two lines left, but the line $y = -3$ is below both points. All you have left is the line $y = 5$, which runs right between the points.

30. (3)

The translation that maps the point $A(1, 2)$ onto $A'(-1, 3)$ subtracts 2 from the x-coordinate (because $1 - 2 = -1$) and adds 1 to the y-coordinate (because $2 + 1 = 3$). You can write the translation like this: $(x, y) \rightarrow (x - 2, y + 1)$.

Under this same translation, the origin $(0, 0)$ is mapped onto $(0 - 2, 0 + 1)$, or $(-2, 1)$.

31. (4)

Solve for x by factoring x out of each term on the left side of the equation:

$$x(x + 5) = 0$$

When the product of two numbers is zero, one of those numbers must be zero. Set each term in the equation to find the solution set:

$$x = 0$$
$$x + 5 = 0; x = -5$$

The solution set is $\{0, -5\}$.

32. (4)

The sum of any two adjacent angles in a parallelogram is 180. Since $m\angle T = 100$, the measure of $\angle MHT$ is 80. Since $\overline{SH}$ bisects $\angle MHT$, it must be true that $m\angle MHS = m\angle SHT = 40$:

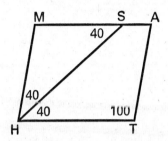

$MATH$ is a parallelogram, so $\overrightarrow{MA}$ is parallel to $\overrightarrow{HT}$. Since $\overrightarrow{SH}$ is a transversal, $\angle MSH$ and $\angle SHT$ are alternate interior angles (which have the same measure). Therefore, $m\angle MSH = 40$. Angles MSH and HSA are supplementary, so their sum is 180. Thus, $m\angle HSA = 140$.

33. (1)

As the answer choices suggest, you can't factor the equation; you have to use the Quadratic Formula:

$$x = \frac{-b \pm \sqrt{b^2 - 4ac}}{2a}$$

In the equation $x^2 + 9x + 12 = 0$, $a = 1$, $b = 9$, and $c = 12$:

$$x = \frac{-9 \pm \sqrt{(9)^2 - 4(1)(12)}}{2(1)} = \frac{-9 \pm \sqrt{81 - 48}}{2} = \frac{-9 \pm \sqrt{33}}{2}$$

34. (2)

When you plot the four points, trapezoid *ABCD* looks like this:

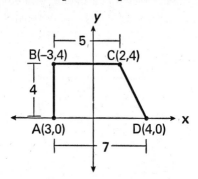

The formula for the area of a trapezoid is:

$$A = \frac{1}{2}(b_1 + b_2)h$$

Use the graph paper to determine the lengths of the two bases: $AD = 7$ and $BC = 5$. The height of the trapezoid can be represented by the length *AB*, which is 4. Plug these numbers into the formula:

$$A = \frac{1}{2}(7 + 5)(4) = \frac{1}{2}(12)(4) = 24$$

35. (3)

Look at the construction; $\angle A'$ has been constructed to be congruent to $\angle A$, and $\angle B'$ has been constructed to be congruent to $\angle B$. The two triangles are therefore similar because of the Angle-Angle Theorem of Similarity. Answer choice (3) restates this rule.

Note: POE is a great help here. Since $\triangle ABC$ and $\triangle A'B'C'$ are clearly not the same size, they're not congruent. Eliminate answer choice (1). Get rid of answer choice (4), because the construction only involves angles, not sides. Be careful of answer choice (2), which is backwards. The triangles are similar because the angles are congruent, not the other way around.

Part II

36. *a* $\dfrac{11y - 30}{10y}$

You can't do anything until the two fractions have the same denominator. Multiply the top and bottom of the first fraction by 5:

$$\frac{5}{5} \cdot \frac{y-4}{2y} = \frac{5(y-4)}{5(2y)} = \frac{5y-20}{10y}$$

Similarly, multiply the top and bottom of the second fraction by 2:

$$\frac{2}{2} \cdot \frac{3y-5}{5y} = \frac{2(3y-5)}{2(5y)} = \frac{6y-10}{10y}$$

Now add them:

$$\frac{5y-20}{10y} + \frac{6y-10}{10y} = \frac{11y-30}{10y}$$

b $\dfrac{-25y^2}{y+2}$

First, turn the division problem into a multiplication problem by flipping the second term:

$$\frac{y^2 - 7y + 10}{5y - y^2} \cdot \frac{25y^3}{y^2 - 4}$$

Now, factor all the complex terms like this:

$$y^2 - 7y + 10 = (y-2)(y-5)$$
$$5y - y^2 = y(5-y)$$
$$y^2 - 4 = (y-2)(y+2)$$

Once you've factored these three terms, the problem looks like this:

$$\frac{(y-2)(y-5)}{y(5-y)} \cdot \frac{25y^3}{(y-2)(y+2)}$$

Cancel out all the factors that appear both on the top and on the bottom, you're left with:

$$\frac{(y-2)(y-5)}{y(5-y)} \cdot \frac{25y^3}{(y-2)(y+2)} = \frac{25y^2(y-5)}{(y+2)(5-y)}$$

You're not done yet. Since $y - 5 = (-1)(5 - y)$, make the substitution:

$$\frac{25y^2(y - 5)}{(y + 2)(5 - y)} = \frac{25y^2(-1)(5 - y)}{(y + 2)(5 - y)} = \frac{-25y^2}{y + 2}$$

37 35

This problem is going to involve a lot of trigonometry, so you should find the measure of some angles. Since $\overline{LK} \cong \overline{LC}$, the angles opposite them, $\angle LCK$ and $\angle K$, are also equal in measure. Therefore, m$\angle LCK = 53$. Use the Rule of 180 on $\triangle AKC$:

$$\text{m}\angle KAC + \text{m}\angle K + \text{m}\angle ACK = 180$$
$$90 + 53 + \text{m}\angle ACK = 180$$
$$\text{m}\angle ACK = 37$$

Since m$\angle ACK$ + m$\angle LCA$ = m$\angle LCK$, you can calculate that m$\angle LCA = 16$:

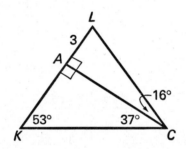

Now do the trig in $\triangle LCA$. You want to find the length LC, which is the hypotenuse of the triangle. You know the length of the opposite leg, $\overline{LA}$. Thus, use the sine (the SOH in SOHCAHTOA):

$$\sin 16° = \frac{LA}{LC}$$

$$0.2756 = \frac{3}{LC}$$

$$0.2756(LC) = 3$$

$$LC = \frac{3}{0.2756}$$

$$LC = 10.88$$

Now you have the lengths of two sides (since $\overline{LK} \cong \overline{LC}$, $LK = 10.88$). There's one more calculation to make, and then you'll be done: you have to find KC. Look at $\triangle ACK$. Since $AL = 3$, $AK = 10.88 - 3$, or 7.88. $\overline{AK}$ is the side adjacent to $\angle K$, and you want to find the hypotenuse KC. Use cosine (the CAH in SOHCAHTOA):

$$\cos 53° = \frac{AK}{KC}$$

$$0.6018 = \frac{7.88}{KC}$$

$$0.6018(KC) = 7.88$$

$$KC = \frac{7.88}{0.6018}$$

$$KC = 13.09$$

Add up all three sides of $\triangle KLC$, and you'll have the perimeter:

$$P = LK + LC + KC = 10.88 + 10.88 + 13.09 = 34.85$$

When you round this off to the nearest meter, as instructed, your answer becomes 35.

38. *a*

Start the graphing process by plugging the integers between 0 and 6, inclusive, into the equation for the parabola and determine its coordinates. For example, if $x = 0$, then $y = -(0)^2 + 6(0) - 8$, or -8. Your first ordered pair is $(0, -8)$. Here is the rest of your T-chart and the accompanying sketch:

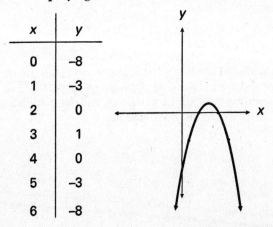

x	y
0	-8
1	-3
2	0
3	1
4	0
5	-3
6	-8

b **1**

From the graph, you can see that the parabola reaches its peak at $x = 3$. Substitute 3 for x in the equation to find the value of y:

$$y = -(3)^2 + 6(3) - 8 = -9 + 18 - 8 = 1$$

Note: If you're not sure where the maximum value of y is, you can determine the axis of symmetry of the parabola using the formula:

$$x = -\frac{b}{2a}$$

In the parabola $y = -x^2 + 6x - 8$, $a = -1$ and $b = 6$:

$$x = -\frac{6}{2(-1)} = \frac{-6}{-2} = 3$$

c **$y = 1$**

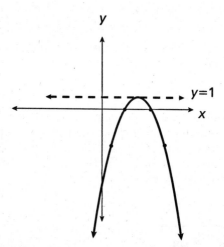

A line parallel to the x-axis is horizontal, so the equation must look like $y = k$ (where k is a constant). Since the point $(3, 1)$ is on the line (it's the turning point of the parabola), and the point has a y-coordinate of 1, the equation of the line is $y = 1$

39. *a* **165**

Since the order in which Elyssa chooses the videos doesn't matter, use the combinations rule:

$$_nC_r = \frac{n!}{r!\,(n-r)!}$$

Elyssa has 11 videos to choose from ($n = 11$), and she wants to choose three ($r = 3$):

$$_{11}C_3 = \frac{11!}{3!\,8!} = \frac{11 \times 10 \times 9 \times 8 \times 7 \times 6 \times 5 \times 4 \times 3 \times 2 \times 1}{3 \times 2 \times 1 \times (8 \times 7 \times 6 \times 5 \times 4 \times 3 \times 2 \times 1)}$$

$$= \frac{990}{6} = 165$$

b **60**

There are four comedies, and Elyssa wants to choose one of them. The number of combinations is $_4C_1$, or 4. There are also six dramas, and Elyssa chooses two of them. Using the same technique, you know that the number of combinations of dramas is $_6C_2$, or 15. The number of three-video combinations consisting of one comedy and two dramas is 4×15, or 60.

c $\dfrac{24}{165}$

First, calculate the number of combinations consisting of one of each type of video. Elyssa can choose one of four comedies ($_4C_1 = 4$), one of six dramas ($_6C_1 = 6$), and only one mystery ($_1C_1 = 1$). The number of combinations is $4 \times 6 \times 1$, or 24. From part A, you know there are 165 possible combinations of three videos, so the probability that a combination will contain one of each type of video is $\dfrac{24}{165}$.

d **6**

In this case, the order of the films does matter, so you have to use the permutations rule:

$$_nP_r = \frac{n!}{r!}$$

There are three videos ($n = 3$), and Elyssa watches one at a time ($r = 1$):

$$_3P_1 = \frac{3!}{1!} = 3 \times 2 \times 1 = 6$$

40. *a*   20 – *x*

From the segment addition postulate, $AD + DC = AC$. Since $AC = 20$ and $AD = x$, you can rewrite the equation as:

$$x + DC = 20$$
$$DC = 20 - x$$

***b*   2**

When you draw the altitude of a right triangle, you cut the triangle into two smaller right triangles. All three of the triangles (the original one, $\triangle ABC$, and the two smaller ones, $\triangle BDC$ and $\triangle ADB$) are similar; all their corresponding sides and angles are proportional to each other.

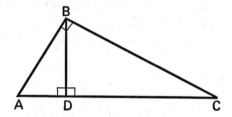

$\overline{AD}$ is the short side of $\triangle ADB$, and $\overline{BD}$ is the long side. In $\triangle BDC$, $\overline{BD}$ is the short side and $\overline{DC}$ is the long side. Set up the proportion:

$$\frac{AD}{BD} = \frac{BD}{DC}$$
$$\frac{x}{6} = \frac{6}{20 - x}$$

Cross-multiply and solve for x:

$$x(20 - x) = 36$$
$$20x - x^2 = 36$$
$$20x - x^2 - 36 = 0$$

Multiply the equation by –1 and rearrange the expression until it's a quadratic in standard form:

$$x^2 - 20x + 36 = 0$$
$$(x - 2)(x - 18) = 0$$
$$x = \{2, 18\}$$

There are two possible values, but the problem specifies that $AD < DC$. Therefore, $x = 2$.

c $2\sqrt{10}$

Use the Pythagorean Theorem on $\triangle ADB$:

$$(AB)^2 = (AD)^2 + (BD)^2$$
$$(AB)^2 = 2^2 + 6^2$$
$$(AB)^2 = 4 + 36$$
$$(AB)^2 = 40$$
$$AB = \sqrt{40}$$

Since the question specifies "simplest radical form," this answer is not sufficient. You have to reduce the radical by factoring out a perfect square:

$$\sqrt{40} = \sqrt{4 \times 10} = \sqrt{4} \times \sqrt{10} = 2\sqrt{10}$$

Part III

41.

The plan: $\overline{BD}$ is an altitude, so it's perpendicular to $\overline{AC}$. Thus, $\angle ADB$ and $\angle CDB$ are right angles. $\overline{BD}$ is also a median, so $\overline{AD} = \overline{DC}$. Prove that $\triangle ADB$ and $\triangle CDB$ are congruent using SAS, then use CPCTC.

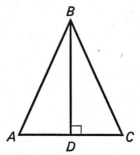

Statements	Reasons
1. $\overline{BD}$ is the median to $\overline{AC}$	1. Given
2. D is the midpoint of $\overline{AC}$	2. Definition of a median
3. $\overline{AD} \cong \overline{DC}$	3. Definition of midpoint
4. $\overline{BD}$ is the altitude to $\overline{AC}$	4. Given
5. $\overline{BD}$ is perpendicular to $\overline{AC}$	5. Definition of an altitude
6. $\angle ADB$ and $\angle CDB$ are right angles	6. Definition of perpendicular
7. $\angle ADB \cong \angle CDB$	7. All right angles are congruent.
8. $\overline{BD} \cong \overline{BD}$	8. Reflexive Property of Congruence
9. $\triangle ADB \cong \triangle CDB$	9. SAS $\cong$ SAS
10. $\overline{BA} \cong \overline{BC}$	10. CPCTC

42. *a*

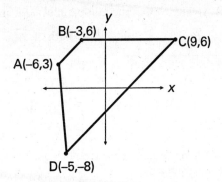

A trapezoid has exactly one pair of opposite parallel sides. To prove that *ABCD* is a trapezoid, you have to show that two of the sides are parallel and the other two are not. To do this, find the slope of each of the sides using the slope formula:

$$m = \frac{y_2 - y_1}{x_2 - x_1}$$

Slope of $\overline{AB}$:   Slope of $\overline{BC}$:   Slope of $\overline{CD}$:   Slope of $\overline{AD}$:

$m = \dfrac{6-3}{-3-(-6)}$ $m = \dfrac{6-6}{9-(-3)}$ $m = \dfrac{-8-6}{-5-9}$ $m = \dfrac{-8-3}{-5-(-6)}$

$= \dfrac{3}{3} = 1$ $= \dfrac{0}{12} = 0$ $= \dfrac{-14}{-14} = 1$ $= \dfrac{-11}{1} = -11$

Since $\overline{AB}$ and $\overline{CD}$ have the same slope, those two sides are parallel. The other two have different slopes, so they are not parallel. Thus, quadrilateral *ABCD* is a trapezoid.

b

In an isosceles trapezoid, the two non-parallel sides have the same length. To prove that *ABCD* is NOT a trapezoid, find the length of the non-parallel sides using the distance formula:

$$d = \sqrt{(x_2 - x_1)^2 + (y_2 - y_1)^2}$$

From part A, you know that the non-parallel sides are $\overline{BC}$ and $\overline{AD}$:

Length of $\overline{BC}$:

$$d = \sqrt{(6-6)^2 + [9-(-3)]^2}$$
$$= \sqrt{0^2 + 12^2}$$
$$= \sqrt{144}$$
$$= 12$$

Length of $\overline{AD}$:

$$d = \sqrt{(-8-3)^2 + [-5-(-6)]^2}$$
$$= \sqrt{(-11)^2 + 1^2}$$
$$= \sqrt{121 + 1}$$
$$= \sqrt{122}$$

Since the two non-parallel sides are not equal in length, $ABCD$ is not an isosceles trapezoid.

EXAMINATION:
JUNE 1997

Part I

Answer 30 questions from this part. Each correct answer will receive 2 credits. No partial credit will be allowed. Write your answers in the spaces provided on the separate answer sheet. Where applicable, answers may be left in terms of π or in radical form. [60]

1 Using the table below, compute $(1 \star 5) \star (2 \star 7)$.

$\star$	1	2	5	7
1	2	7	1	5
2	7	5	2	1
5	1	2	5	7
7	5	1	7	2

2 In the accompanying diagram, line l is parallel to line k, line $m \perp$ line k, and $m\angle x = m\angle y$. Find $m\angle x$.

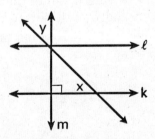

3 If ♥ is a binary operation defined as $a ♥ b = \sqrt{a^2 + b^2}$, find the value of 12 ♥ 5.

4 In the accompanying diagram of similar triangles ABE and ACD, $\overline{ABC}$, $\overline{AED}$, $AB = 6$, $BC = 3$, and $ED = 4$. Find the length of $\overline{AE}$.

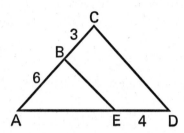

5 How many different 5-letter arrangements can be formed from the letters in the word "DANNY"?

6 Evaluate: $\dfrac{9!}{3!\,5!}$

7 In the accompanying diagram of $\triangle ABC$, $\overline{AB}$ is extended to E and D, exterior angle CBD measures $130°$, and m$\angle C = 75$. Find m$\angle CAE$.

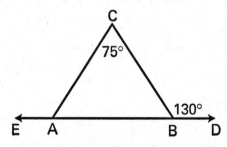

8 In right triangle ABC, $\angle C$ is a right triangle and m$\angle B$ = 60. What is the ratio of m$\angle A$ to m$\angle B$?

9 In $\triangle ABC$, m$\angle A$ = $3x + 40$, m$\angle B$ = $8x + 35$, and m$\angle C$ = $10x$. Which is the longest side of the triangle?

10 A bookshelf contains seven math textbooks and three science textbooks. If two textbooks are drawn at random without replacement, what is the probability both books are science textbooks?

11 Express the product in lowest terms:

$$\frac{x^2 - x - 6}{3x - 9} \cdot \frac{2}{x + 2}$$

12 In rhombus $ABCD$, the measure of $\angle A$ is 30° more than twice the measure of $\angle B$. Find m$\angle B$.

13 The endpoints of the diameter of a circle are $(-6,2)$ and $(10,-2)$. What are the coordinates of the center of the circle?

14 Find the area of a triangle whose vertices are $(1,2)$, $(8,2)$, and $(1,6)$.

15 Find the distance between points $(-1,1)$ and $(2,-5)$.

16   In the accompanying diagram, the bisectors of ∠A
     and ∠B in acute triangle ABC meet at D, and
     m∠ADB = 130. Find m∠C.

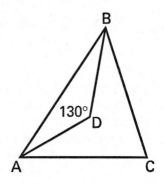

17   Point P is on line m. What is the total number of
     points 3 centimeters from line m and 5
     centimeters from point P?

18   The diagonals of a rhombus are 8 and 10. Find
     the measure of a side of the rhombus to the
     nearest tenth.

Directions (19-34): For *each* question chosen, write the
numeral preceding the word or expression that best completes
the statement or answers the question.

19   In isosceles triangle ABC, $\overline{AB} \cong \overline{BC}$, point D lies
     on $\overline{AC}$, and $\overline{BD}$ is drawn. Which inequality is true?
     (1) m∠A > m∠ADB   (3) BD > AB
     (2) m∠C > m∠CDB   (4) AB > BD

20 If the statements m, $m \rightarrow p$, and $r \rightarrow \sim p$ are true, which statement must also be true?

(1) $\sim r$ (3) $r \wedge \sim p$
(2) $\sim p$ (4) $\sim p \vee \sim m$

21 If a point in Quadrant IV is reflected in the y-axis, its image will lie in Quadrant

(1) I   (3) III
(2) II   (4) IV

22 In right triangle ABC, m$\angle C$ = 90, m$\angle A$ = 63, and AB = 10. If BC is represented by a, then which equation can be used to find a?

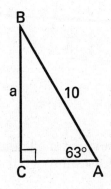

(1) $\sin 63^\circ = \dfrac{a}{10}$ (3) $\tan 63^\circ = \dfrac{a}{10}$
(2) $a = 10 \cos 63^\circ$ (4) $a = \tan 27^\circ$

23 If point $R'(6,3)$ is the image of point $R(2,1)$ under a dilation with respect to the origin, what is the constant of the dilation?

(1) 1   (3) 3
(2) 2   (4) 6

24 What is an equation of a line that passes through the point $(0,3)$ and is perpendicular to the line whose equation is $y = 2x - 1$?

(1) $y = -2x + 3$

(3) $y = -\dfrac{1}{2}x + 3$

(2) $y = 2x + 3$

(4) $y = \dfrac{1}{2}x + 3$

25 What is an equation of the function shown in the accompanying diagram?

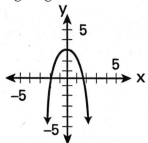

(1) $y = x^2 + 3$

(3) $y = -x^2 - 3$

(2) $y = -x^2 + 3$

(4) $y = (x^2 - 3)^2$

26 What is an equation of the line that is parallel to the y-axis and passes through the point $(2,4)$?

(1) $x = 2$

(3) $x = 4$

(2) $y = 2$

(4) $y = 4$

27 In the accompanying diagram, the altitude to the hypotenuse of right triangle ABC is 8.

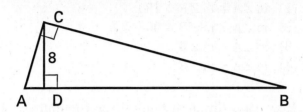

The altitude divides the hypotenuse into segments whose measures may be

(1) 8 and 12
(2) 3 and 24
(3) 6 and 10
(4) 2 and 32

28 If the coordinates of the center of a circle are $(-3,1)$ and the radius is 4, what is an equation of the circle?

(1) $(x - 3)^2 + (y + 1)^2 = 4$
(2) $(x + 3)^2 + (y - 1)^2 = 16$
(3) $(x + 3)^2 + (y - 1)^2 = 4$
(4) $(x - 3)^2 + (y + 1)^2 = 16$

29 Which expression is a solution for the equation $2x^2 - x = 7$?

(1) $\dfrac{-1 \pm \sqrt{57}}{2}$

(2) $\dfrac{1 \pm \sqrt{57}}{2}$

(3) $\dfrac{-1 \pm \sqrt{57}}{4}$

(4) $\dfrac{1 \pm \sqrt{57}}{4}$

30 If the complement of $\angle A$ is greater than the supplement of $\angle B$, which statement *must* be true?

(1) $m\angle A + m\angle B = 180$
(2) $m\angle A + m\angle B = 90$
(3) $m\angle A < m\angle B$
(4) $m\angle A > m\angle B$

31 How many different four-person committees can be formed from a group of six boys and four girls?

(1) $\dfrac{10!}{4!}$

(2) $_{10}P_4$

(3) $_6C_2 \bullet {}_4C_2$

(4) $_{10}C_4$

32 Which equation represents the axis of symmetry of the graph of the eqaution $y = x^2 - 4x - 12$?

(1) $y = 4$
(2) $x = 2$
(3) $y = -2$
(4) $x = -4$

33 What is $\dfrac{1}{x} + \dfrac{1}{1-x}$, $x \neq 1,0$, expressed as a single fraction?

(1) $\dfrac{1}{x(1-x)}$

(2) $\dfrac{-1}{x(x+1)}$

(3) $\dfrac{2}{-x}$

(4) $\dfrac{1}{x(x-1)}$

34 In the accompanying diagram, $\overline{RL} \perp \overline{LP}$, $\overline{LR} \perp \overline{RT}$, and M is the midpoint of $\overline{TP}$.

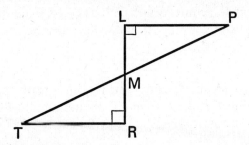

Which method could be used to prove $\triangle TMR \cong \triangle PML$?

(1) SAS ≅ SAS   (3) HL ≅ HL
(2) AAS ≅ AAS   (4) SSS ≅ SSS

Directions (35): Leave all construction lines on the answer sheet.

35 On the answer sheet, construct an equilateral triangle in which $\overline{AB}$ is one of the sides.

A   B

Part Two

Answer *three* questions from this part. Clearly indicate the necessary steps, including appropriate formula substitutions, diagrams, graphs, charts, etc. Calculations that may be obtained by mental arithmetic or the calculator do not need to be shown. [40]

36 *a* On graph paper, draw the graph of the equation $y = x^2 - 8x + 2$, including all values of x in the interval $0 \leq x \leq 8$. [6]

b Find the roots of the equation $x^2 - 8x + 2 = 0$ to the *nearest hundredth*. [*Only an algebraic solution will be accepted.*] [4]

37 The coordinates of the endpoints of $\overline{AB}$ are $A(-2,4)$ and $B(4,1)$.

a On a set of axes, graph $\overline{AB}$. [1]

b On the same set of axes, graph and state the coordinates of

(1) $\overline{A'B'}$, the image of $\overline{AB}$ after a reflection in the x-axis [2]

(2) $\overline{A''B''}$, the image of $\overline{A'B'}$ after a translation that shifts (x,y) to $(x + 2,y)$ [2]

c Using coordinate geometry, determine if $\overline{A'B'} \cong \overline{A''B''}$. Justify your answer. [5]

38 Answer both *a* and *b* for all values for which these expressions are defined.

a Solve for *x*: $-\dfrac{2}{5} + \dfrac{x+4}{x} = 1$ [4]

b Express the difference in simplest form:

$$\frac{3y}{y^2 - 4} - \frac{2}{y + 2}$$ [6]

39 Solve the following system of equations algebraically and check:

$$y = 2x^2 - 4x - 5$$
$$2x + y + 1 = 0$$ [8,2]

40 In the accompanying diagram of $\triangle ABC$, altitude $AD = 13$, $\overline{AB} \cong \overline{AC}$, and m$\angle BAC = 70$.

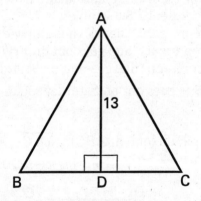

a Find *BC* to the *nearest tenth*. [4]

b Using the answer from part *a*, find, to the *nearest tenth*, the

(1) area of $\triangle ABC$ [2]

(2) perimeter of $\triangle ABC$ [4]

Part III

Answer *one* question from this part. Clearly indicate the necessary steps, including appropriate formula substitutions, diagrams, graphs, charts, etc. Calculations that may be obtained by mental arithmetic or the calculator do not need to be shown. [10]

41 Given: If Sue goes out on Friday night and not on Saturday night, then she does not study.

If Sue does not fail mathematics, then she studies.

Sue does not fail mathematics.

If Sue does not go out on Friday night, then she watches a movie.

Sue does not watch a movie.

Let *A* represent: "Sue fails mathematics."
Let *B* represent: "Sue studies."
Let *C* represent: "Sue watches a movie."
Let *D* represent: "Sue goes out on Friday night."
Let *E* represent: "Sue goes out on Saturday night."

Prove: Sue goes out on Saturday night. [10]

42 Given: parallelogram $ABCD$, $\overline{DFC}$, $\overline{AEB}$, $\overline{ED}$ bisects $\angle ADC$, and $\overline{FB}$ bisects $\angle ABC$.

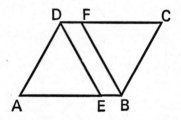

Prove: $\overline{EB} \cong \overline{DF}$ [10]

ANSWER KEY

1. 2	13. (2,0)	25. (2)
2. 45	14. 14	26. (1)
3. 13	15. 5	27. (4)
4. 8	16. 80	28. (2)
5. 60	17. 4	29. (4)
6. 504	18. 6.4	30. (3)
7. 125	19. (4)	31. (4)
8. $\frac{1}{2}$	20. (1)	32. (2)
9. $\overline{AC}$	21. (3)	33. (1)
10. $\frac{1}{15}$	22. (1)	34. (2)
11. $\frac{2}{3}$	23. (3)	35. construction
12. 50	24. (3)	

EXPLANATIONS: JUNE 1997

Part I

1. 2

Figure out the terms in parentheses first. To calculate 1 ✳ 5, find the 1 in the leftmost column and run your finger along that row until you to the column headed by the 5. You'll find that 1 ✳ 5 = 1:

✳	1	2	5	7
1	2	7	1	5
2	7	5	2	1
5	1	2	5	7
7	5	1	7	2

Now find the other term in parentheses: 2 ✳ 7 = 1. The problem now looks like this:

$$(1 ✳ 5) ✳ (2 ✳ 7) = 1 ✳ 1 = 2$$

2. 45

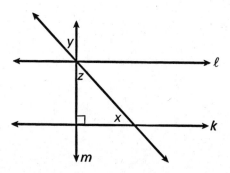

Look at the triangle in the center of the diagram: The sum of the three angles is 180°. Since one of the angles is a right angle, the sum of the other two angles, $\angle x$ and $\angle z$ is 90. Now look at the angles. The problem states that m $\angle x$ = m $\angle y$. Since $\angle y$ and $\angle z$ are vertical angles, m $\angle y$ = m $\angle z$. Therefore, m $\angle x$ = m $\angle z$.

Since angles x and z are equal to each other and have a sum of 90, each angle must measure 45°. Therefore, m$\angle x$ = 45.

3. 13

Don't let the symbols freak you out. This function question defines what the "♥" means, so all you have to do is plug in $a = 12$ and $b = 5$:

$$a ♥ b = \sqrt{a^2 + b^2}$$
$$12 ♥ 5 = \sqrt{12^2 + 5^2} = \sqrt{144 + 25} = \sqrt{169} = 13$$

4. 8

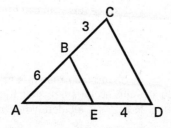

Corresponding sides of the two triangles are proportional. Since $AC = AB + BC$, you can infer that $AC = 9$. Let $AE = x$ and set up the following proportion:

$$\frac{AB}{AC} = \frac{AE}{AD}$$
$$\frac{6}{9} = \frac{x}{x + 4}$$

Now cross-multiply:

$$9x = 6(x + 4)$$
$$9x = 6x + 24$$
$$3x = 24$$
$$x = 8$$

5. 60

The formula to follow is a variation of the permutations rule. To find the number of possible arrangements of the letter in a word with n letters, in which one letter appears p times (when that p is greater than 1), the formula looks like this:

$$\frac{n!}{p!}$$

DANNY has 5 letters, but there are two N's. Therefore, you can express the number of arrangements as:

$$\frac{5!}{2!} = \frac{5 \times 4 \times 3 \times 2 \times 1}{2 \times 1} = 60$$

If your name is Danny, think of how you can impress your friends at parties!

6. 504

Wow. Two problems involving factorials in a row. The term $n!$ represents the product of all integers between 1 and n, inclusive. For example, $4! = 4 \times 3 \times 2 \times 1$. Calculate your answer like this:

$$\frac{9!}{3!\,5!} = \frac{9 \times 8 \times 7 \times 6 \times 5 \times 4 \times 3 \times 2 \times 1}{3 \times 2 \times 1 \times (5 \times 4 \times 3 \times 2 \times 1)} = 9 \times 8 \times 7 = 504$$

7. 125

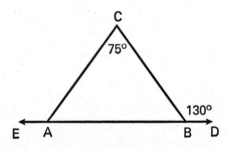

The measure of an exterior angle of a triangle equals the sum of the two non-adjacent interior angles, and $\angle CBD$ is an exterior angle:

$$m\angle CAB + m\angle C = m\angle CBD$$
$$m\angle CAB + 75 = 130$$
$$m\angle CAB = 55$$

Since $\angle CAE$ and $\angle CAB$ are supplementary, $m\angle CAE = 125$.

8. $\dfrac{1}{2}$

Given that $\angle C$ is a right angle, m $\angle C$ = 90. Since m $\angle B$ = 60, the Rule of 180 dictates that m $\angle A$ = 30.

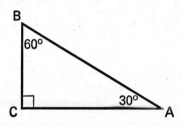

The ratio of m $\angle A$ to m $\angle B$ is $\dfrac{30}{60}$, or $\dfrac{1}{2}$.

9. $\overline{AC}$

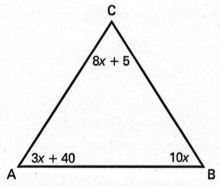

Solve for x first, using the Rule of 180:

$$\text{m} \angle A + \text{m} \angle B + \text{m} \angle C = 180$$
$$(3x + 40) + (8x + 35) + 10x = 180$$
$$21x + 75 = 180$$
$$21x = 105$$
$$x = 5$$

Now find the size of each angle:

$$m \angle A = 3(5) + 40 = 55$$
$$m \angle B = 8(5) + 35 = 75$$
$$m \angle C = 10(5) = 50$$

Since $\angle B$ is the largest angle of $\triangle ABC$, the side opposite $\angle B$, side $\overline{AC}$, is the largest side.

10. $\dfrac{1}{15}$

There is a total of 10 textbooks on the shelf, and three of them are science textbooks. Therefore, the probability that you'll select a science textbook the first time is $\dfrac{3}{10}$.

After one book is selected, there are nine books left and only two of them are science textbooks. The probability that a second book is selected is $\dfrac{2}{9}$. To find the probability that the first two books selected are science textbooks, multiply these two fractions together:

$$\frac{3}{10} \times \frac{2}{9} = \frac{6}{90}$$

This fraction reduces to $\dfrac{1}{15}$.

11. $\dfrac{2}{3}$

If you do a little factoring, you'll see how nicely this thing reduces:

$$x^2 - x - 6 = (x - 3)(x + 2)$$
$$3x - 9 = 3(x - 3)$$

The product now looks like this:

$$\frac{(x - 3)(x + 2)}{3(x - 3)} \cdot \frac{2}{x + 2}$$

Cross off the terms that appear both on the top and bottom, and you're left with:

$$\frac{(x - 3)(x + 2)}{3(x - 3)} \cdot \frac{2}{x + 2} = \frac{2}{3}$$

12. 50

If m $\angle B = x$, then m $\angle A = 2x + 30$:

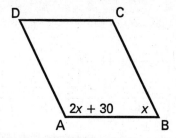

All rhombi are parallelograms, and the sum of any two consecutive angles in a parallelogram is 180°. Since $\angle A$ and $\angle B$ are consecutive angles, m $\angle A$ + m $\angle B = 180$:

$$x + (2x + 30) = 180$$
$$3x + 30 = 180$$
$$3x = 150$$
$$x = 50$$

Check your work when you're done: if m $\angle B = 50$, then m $\angle A = 2(50) + 30$, or 130. Since 50 + 130 = 180, the two angles are supplementary.

13. (2, 0)

If the endpoints of the diameter of a circle are (–6, 2) and (10, –2), then the center of the circle must be the midpoint of the segment between those two points:

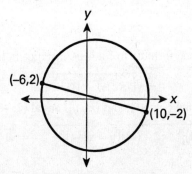

The formula for the midpoint of a segment is:

$$(\bar{x}, \bar{y}) = \left(\frac{x_1 + x_2}{2}, \frac{y_1 + y_2}{2} \right)$$

Therefore, the midpoint of the circle's diameter is·

$$(\bar{x}, \bar{y}) = \left(\frac{x_1 + x_2}{2}, \frac{y_1 + y_2}{2} \right)$$

$$(\bar{x}, \bar{y}) = \left(\frac{-6 + 10}{2}, \frac{2 + (-2)}{2} \right) = \left(\frac{4}{2}, \frac{0}{2} \right) = (2, 0)$$

14. 14

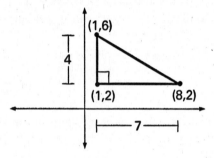

From the diagram, you can discern that the triangle is a right triangle. The base of the triangle is 7 units long ($b = 7$), and the triangle's height is 4 units ($h = 4$). Use the formula for the area of a triangle to find the triangle's area:

15. 5

Use the distance formula to find the distance between two points:

$$d = \sqrt{(y_2 - y_1)^2 + (x_2 - x_1)^2}$$

Let (x_1, y_1) equal $(-1, -1)$, and (x_2, y_2) equal $(2, -5)$:

$$d = \sqrt{[-5 - (-1)]^2 + [2 - (-1)]^2} = \sqrt{(-4)^2 + 3^2} = \sqrt{16 + 9} = \sqrt{25} = 5$$

16. 80

This one's tricky because it requires a little imagination. Look at △*ADB*. The sum of all three angles in △*ADB* equals 180, so m∠*BAD* + m∠*DBA* = 50.

Here's where the imagination kicks in. There's no way you can know the exact measures of ∠*BAD* and ∠*DBA*, but you don't have to. Their sum is 50, so let m∠*BAD* = 30 and m∠*DBA* = 20:

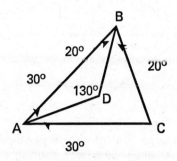

Since $\overline{AD}$ bisects ∠*BAC*, m∠*BAC* = 60. Similarly, m∠*CBA* = 40. Now you can use the Rule of 180 to find m∠*C*:

$$m\angle BAC + m\angle CBA + m\angle C = 180$$
$$60 + 40 + m\angle C = 180$$
$$m\angle C = 80$$

17. 4

The locus of points 5 centimeters from point *P* is a circle with a radius of 5. The locus of points that are 3 centimeters from line *m* is two lines. The diagram looks like this:

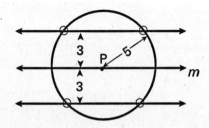

There are four points of intersection.

18. 6.4

The diagonals of a rhombus bisect each other. Since the length of diagonal $\overline{AC}$ is 10, then $AE = 5$. Using this same logic, you can figure out that $BE = 4$.

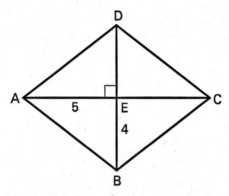

The diagonals of a rhombus are also perpendicular to each other, so $\triangle ABE$ is a right triangle. Therefore, you can use the Pythagorean Theorem to find the length of $\overline{AB}$:

$$(AB)^2 = (AE)^2 + (BE)^2$$
$$(AB)^2 = 5^2 + 4^2$$
$$(AB)^2 = 25 + 16$$
$$(AB)^2 = 41$$
$$AB = \sqrt{41}$$

Use your calculator to find $\sqrt{41}$, which equals 6.403. When you round this to the nearest tenth, as instructed, you get 6.4.

Multiple Choice

19. (4)

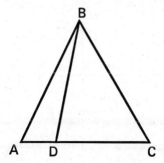

Since $AB = BC$, any line segment drawn from B to $\overline{AC}$ has to be shorter than either side of the triangle. Therefore, AB has to be bigger than BD.

20. (1)

Here's a little mini-proof. Given that m and $m \rightarrow p$, the Law of Detachment dictates that the result is p. It looks like this in symbolic form:

$$[(m \rightarrow p) \land m] \rightarrow p$$

The second rule to use is the Rule of *Modus Tollens*, which combines the Law of Detachment and the Law of Contrapositive Inference:

$$[(r \rightarrow \sim p) \land p] \rightarrow \sim r$$

The statement $\sim r$ must be true.

21. (3)

The quadrants look like this:

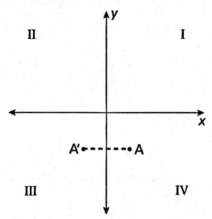

If point *A* in Quadrant IV is reflected in the *y*-axis, the image of the point, *A′*, will appear in Quadrant III.

22. (1)

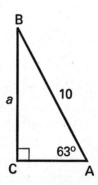

Side *a* is the leg opposite ∠*A*, and $\overline{AB}$ is the hypotenuse, which is 10 units long. Since you know the opposite and the hypotenuse, use the sine (the SOH in SOHCAHTOA):

$$\sin \angle A = \frac{\text{opposite}}{\text{hypoteneuse}}$$

$$\sin 63° = \frac{a}{10}$$

Since no other answer choice mentions the sine, you can cross them all off.

23. (3)

When a point undergoes a dilation, each coordinate of that point is multiplied by a constant. The image of (x, y), for example, is (kx, ky).

Each coordinate of point $R(2, 1)$ is multiplied by 3 in order to map it onto its image, $R(6, 3)$. Therefore, the constant of the dilation is 3.

24. (3)

Because the equation $y = 2x - 1$ is in slope-intercept form ($y = mx + b$), you know that the slope of the line is 2. The slopes of perpendicular lines are negative reciprocals (their product is -1), so the slope of the line you're looking for is $-\frac{1}{2}$.

Each of the answer choices is also in slope-intercept form, and the only one with a slope of $-\frac{1}{2}$ is answer choice (3).

25. (2)

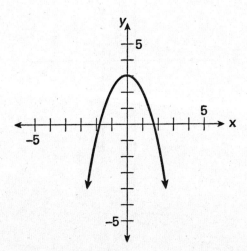

The parabola in this question opens down, so the coefficient of the x^2 term, otherwise known as a, is negative. (The rule is: if $a > 0$, the parabola "smiles"; if $a < 0$, it "frowns.") Eliminate answer choices (1) and (4), because a is positive in each of them.

The only bit of information left is the y-intercept. The parabola intercepts the y-axis at the point $(0, 3)$, so its y-intercept is 3. Therefore, the proper equation is $y = -x^2 + 3$.

Note: Even though there are only two terms in the equation, the equation of this parabola is still in standard form. It just happens that $b = 0$.

26. (1)

Lines that are parallel to the y-axis are vertical and have the equation $x = h$, in which h is a constant. You can eliminate answer choices (2) and (4), because they have y's in them.

Now look at the point in the question. The x-coordinate of $(2, 4)$ is 2, so the equation of the line must be $x = 2$.

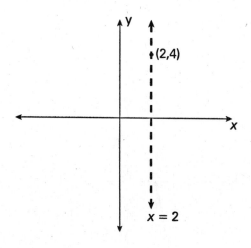

27. (4)

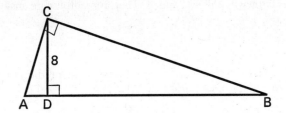

When you draw the altitude of a right triangle, you cut the triangle into two smaller right triangles. All three of the triangles (the original one, $\triangle ABC$, and the two smaller ones, $\triangle ACD$ and $\triangle CBD$) are similar; all their corresponding sides are proportional to each other. In $\triangle ACD$, AD is the short leg and CD is the long leg. In $\triangle CBD$, CD is the short leg and DB is the long leg. Therefore:

$$\frac{AD}{CD} = \frac{CD}{DB}$$

Since $CD = 8$, you can substitute and cross-multiply:

$$\frac{AD}{8} = \frac{8}{DB}$$
$$AD \times DB = 64$$

The product of the lengths AD and DB must be 64, and only answer choice (4) fulfills this requirement. To be sure, you can check the proportion:

$$\frac{2}{8} = \frac{8}{32}$$
$$\frac{1}{4} = \frac{1}{4}$$

28. (2)

Use the formula for a circle, and remember that (h, k) is the center and r is the radius:

$$(x - h)^2 + (y - k)^2 = r^2$$
$$[x - (-3)]^2 + (y - 1)^2 = 4^2$$
$$(x + 3)^2 + (y - 1)^2 = 16$$

Since the formula involves r^2 and not r, you should recognize that the formula will equal 16, not 4. Therefore, eliminate answer choices (1) and (3).

29. (4)

As the answer choices suggest, you can't factor the equation; you have to use the Quadratic Formula:

$$x = \frac{-b \pm \sqrt{b^2 - 4ac}}{2a}$$

Before you do that, though, you have to put the equation in standard form by subtracting 7 from both sides:

$$2x^2 - x = 7$$
$$2x^2 - x - 7 = 0$$

In the equation $2x^2 - x - 7 = 0$, $a = 2$, $b = -1$, and $c = -7$:

$$x = \frac{-(-1) \pm \sqrt{(-1)^2 - 4(2)(-7)}}{2(2)} = \frac{1 \pm \sqrt{1 + 56}}{4} = \frac{1 \pm \sqrt{1 + 57}}{4}$$

30. (3)

To get a grip on this problem, plug in some numbers (and make sure the numbers you choose fit the requirements). If m $\angle A$ = 40, then the complement of $\angle A$ is 50. If m $\angle B$ = 160, then the supplement of $\angle B$ is 20. At this point, you can tell that m $\angle A$ < m $\angle B$, and that the rest of the answer choices can't be correct.

31. (4)

Don't be distracted by the fact that there are six boys and four girls. It doesn't matter how many boys and girls there are in the four-person group, so you only have to note that there are 10 kids to choose from.

To find the number of ways you can choose four of 10 people, use the combinations rule:

$$_{10}C_4$$

You don't even have to solve it.

32. (2)

When a parabola appears in the form $y = ax^2 + bx + c$ as this one does, the equation of its axis of symmetry is:

$$x = -\frac{b}{2a}$$

For this parabola, $a = 1$, $b = -4$, and $c = -12$. Thus the axis of symmetry is:

$$x = -\frac{-4}{2(1)} = \frac{4}{2} = 2$$

The equation of the axis of symmetry is $x = 2$.

Note: When the equation of a parabola is in standard form, its axis of symmetry is vertical and its equation is $x = k$, in which k is a constant. Therefore, you can eliminate answer choices (1) and (3).

33. (1)

You can't do anything until the two fractions have the same denominator. To make two fractions compatible, multiply the top and bottom of the first fraction by $(1 - x)$:

$$\frac{(1-x)}{(1-x)} \cdot \frac{1}{x} = \frac{1-x}{x(1-x)}$$

Multiply the top and bottom of the second fraction by x:

$$\frac{x}{x} \cdot \frac{1}{1-x} = \frac{x}{x(1-x)}$$

Now you can add the fractions:

$$\frac{1-x}{x(1-x)} + \frac{x}{x(1-x)} = \frac{1-x+x}{x(1-x)} = \frac{1}{x(1-x)}$$

34. (2)

Be careful here; don't be suckered into choosing Hy-Leg just because the triangles in the diagram are right triangles. You know that the hypotenuses are congruent because M is the midpoint of $\overline{TP}$, but you don't know anything about either of the legs.

All right angles are congruent, so $\angle L \cong \angle R$. Further, $\angle LMP$ and $\angle TMR$ are vertical angles, so they're congruent. Since you know of two congruent angles, the congruency theorem has two A's in it. The only one that does is answer choice (2).

35. construction

This is one of the easiest constructions to do (if you remember it, that is). Make your compass exactly as wide as $\overline{AB}$; put the pointy end on A and make an arc above the segment, like this:

Without changing the width of your compass, put the point on B and make another arc above the segment. Identify the point of intersection as point C.

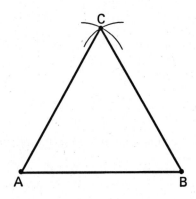

Once you draw $\overline{AC}$ and $\overline{BC}$, you've constructed equilateral triangle ABC.

Part II

36. *a*

Start the graphing process by plugging the integers between 0 and 8, inclusive, into the equation for the parabola and determine its coordinates. For example, if $x = 0$, then $y = (0)^2 - 8(0) + 2$, or 2. Your first ordered pair is (0, 2). Here's the rest of your T-chart and the accompanying sketch:

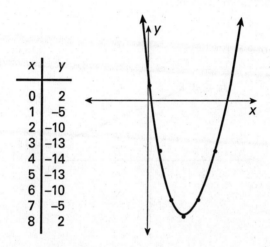

x	y
0	2
1	−5
2	−10
3	−13
4	−14
5	−13
6	−10
7	−5
8	2

b $x = \{7.74, 0.26\}$

You can't factor the equation (the fact that you have to solve for x to the nearest hundredth should give you a hint), so you have to use the Quadratic Formula:

$$x = \frac{-b \pm \sqrt{b^2 - 4ac}}{2a}$$

In the equation $x^2 - 8x + 2 = 0$, $a = 1$, $b = -8$, and $c = 2$:

$$x = \frac{-(-8) \pm \sqrt{(-8)^2 - 4(1)(2)}}{2(1)} = \frac{8 \pm \sqrt{64 - 8}}{2} = \frac{8 \pm \sqrt{56}}{2}, \frac{8 - \sqrt{56}}{2}$$

Use your calculator to find that $\sqrt{56} = 7.48$ and substitute:

$$x = \frac{8 + 7.48}{2} \qquad\qquad x = \frac{8 + 7.48}{2}$$

$$= \frac{15.48}{2} \qquad\qquad\qquad = \frac{0.52}{2}$$

$$-7.74 \qquad\qquad\qquad\qquad = 0.26$$

37. a

The graph is the easy part:

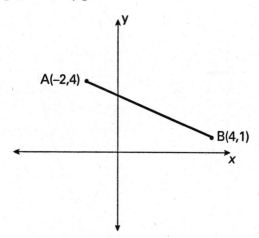

b (1) A′(−2, −4), B′(4, −1)

After a reflection in the x-axis, the y-coordinate of each point is negated and the x-coordinate of each point remains unchanged. In other words, $r_{x\text{-axis}}\,(x, y) \to (x, -y)$.

Therefore, the image of $A(−2, 4)$ is $A′(−2, −4)$, and the image of $B(4, 1)$ is $B′(4, −1)$.

(2) A″(0, −4), B″(6, −1)

Under a translation $(x + 2, y)$, each x-coordinate increases by 2 and each y-coordinate remains the same. The image of $A′(−2, 4)$ is $A″(−2 + 2, 4)$, or $A″(0, 4)$; the image of $B′(4, −1)$ is $B″(4 + 2, 1)$, or $B″(6, 1)$.

Your graph should now look like this:

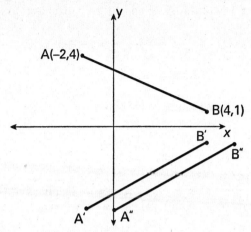

c

It should be obvious that $A'B' = A''B''$, because a translation doesn't change the size of an object. The Regents folks want you to prove it using algebra, so use the distance formula to find the length of each segment:

$$d = \sqrt{(x_2 - x_1)^2 + (y_2 - y_1)^2}$$

$$A'B' = \sqrt{[4 - (-2)]^2 + [-1 - (-4)]^2} \qquad A''B'' = \sqrt{(6 - 0)^2 + [-1 - (-4)]^2}$$

$$= \sqrt{6^2 + 3^2} \qquad\qquad\qquad = \sqrt{6^2 + 3^2}$$

$$= \sqrt{36 + 9} \qquad\qquad\qquad = \sqrt{36 + 9}$$

$$= \sqrt{45} \qquad\qquad\qquad\quad = \sqrt{45}$$

$$= 3\sqrt{5} \qquad\qquad\qquad\quad = 3\sqrt{5}$$

The two segments have the same length.

38. *a* 10

The easiest way to start is to add $\frac{2}{5}$ to both sides:

$$-\frac{2}{5} + \frac{x+4}{x} + \frac{2}{5} = 1 + \frac{2}{5}$$

$$\frac{x+4}{x} = \frac{7}{5}$$

Now you can cross-multiply and solve:

$$7x = 5(x + 4)$$
$$7x = 5x + 20$$
$$2x = 20$$
$$x = 10$$

As always, remember to check your work by plugging 10 in for x in the original equation and making sure it works.

b $\dfrac{y+4}{y^2-4}$

You can't do anything until the two fractions have the same denominator. Since $y^2 - 4 = (y + 2)(y - 2)$, you don't have to do anything to the first fraction. Multiply the top and bottom of the second fraction by $(y - 2)$:

$$\frac{(y-2)}{(y-2)} \cdot \frac{2}{y+2} = \frac{2(y-2)}{(y+2)(y-2)} = \frac{2y-4}{y^2-4}$$

Now subtract them:

$$\frac{3y}{y^2-4} - \frac{2y-4}{y^2-4} = \frac{3y - (2y-4)}{y^2-4} = \frac{y+4}{y^2-4}$$

Be careful with the minus signs. You're subtracting $2y - 4$ from $3y$, so be sure to use parentheses. Otherwise, your numerator will be $y - 4$, and you'll lose points.

39. (2, –5) and (–1, 1)

Look at the second equation first, and rearrange it until only y appears on the left side of the equation:

$$2x + y + 1 = 0$$
$$2x + y = -1$$
$$y = -2x - 1$$

Set the two equations equal to each other like this:

$$y = 2x^2 - 4x - 5$$
$$y = -2x - 1$$
$$2x^2 - 4x - 5 = -2x - 1$$
$$2x^2 - 2x - 5 = -1$$
$$2x^2 - 2x - 4 = 0$$

Since each coefficient is divisible by 2, divide each term to make the quadratic easier to factor:

$$x^2 - x - 2 = 0$$

Factor and solve for x:

$$(x - 2)(x + 1) = 0$$
$$x = \{2, -1\}$$

You have the two x-values, so you have to find each corresponding y-values:

If $x = 2$, then $y = 2(2)^2 - 4(2) - 5$, or -5. The first solution is $(2, -5)$.

If $x = -1$, then $y = 2(-1)^2 - 4(-1) - 5$, or 1. The second solution is $(-1, 1)$.

Be sure to check your work by substituting back into the equations, or you'll lose two points.

40. *a* 18.2

Since $\overline{AB} \cong \overline{AC}$, $\triangle ABC$ is isosceles. The altitude from the vertex of an isosceles triangle bisects the vertex angle, so m $\angle BAD$ = m $\angle DAC$ = 35.

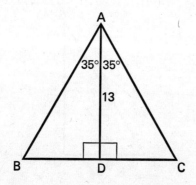

Use trigonometry in $\triangle ABD$ to find BD. You know the adjacent leg ($AD = 13$), and you want to find the opposite leg. Use tangent (the TOA in SOHCAHTOA):

$$\tan \angle BAD = \frac{BD}{AD}$$

$$\tan 35° = \frac{BD}{13}$$
$$13(0.7002) = BD$$
$$9.1 = BD$$

The altitude of an isosceles triangle is also a median of that triangle. Therefore, D is the midpoint of $\overline{AC}$ and $BD = DC$. Thus, BC is twice as long as BD, or 18.2.

b **(1) 118.3**

The area is a piece of cake. You know the length of the base ($BC = 18.2$) and the height ($AD = 13$), so use the formula for the area of a triangle:

$$A = \frac{1}{2}bh = \frac{1}{2}(18.2)(13) = 118.3$$

(2) 49.9

You know that $AB = AC$, so find the length of $\overline{AB}$ using the Pythagorean Theorem. From part A, you know that $BD = 9.1$:

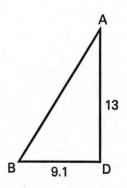

$$(AB)^2 = (AD)^2 + (BD)^2$$
$$(AB)^2 = 13^2 + (9.1)^2$$
$$(AB)^2 = 169 + 82.81$$
$$(AB)^2 = 251.81$$
$$AB = \sqrt{251.81} = 15.87$$

Add up all three sides of $\triangle ABC$, and you'll have the perimeter:

$$P = AB + BC + AC = 15.87 + 18.2 + 15.87 = 49.94$$

When you round this off to the nearest *tenth* of a meter, as instructed, your answer becomes 49.9.

Part III

41.

In this proof, they've already given you the symbols.

Step One: Turn all the givens into symbolic terms:

"If Sue goes out on Friday night and not on Saturday night, then she does not study." $(D \wedge {\sim}E) \rightarrow {\sim}B$

"If Sue does not fail mathematics, then she studies." ${\sim}A \rightarrow B$

"Sue does not fail mathematics." ${\sim}A$

"If Sue does not go out on Friday night, then she watches a movie." ${\sim}D \rightarrow C$

"Sue does not watch a movie." ${\sim}C$

Step Two: Decide what you want to prove:

"We will not buy souvenirs." E

Step Three: Write the proof.

Statements	Reasons
1. $\sim A \rightarrow B$ $\sim A$	1. Given
2. B	2. Law of Detachment
3. $(D \wedge \sim E) \rightarrow \sim B$	3. Given
4. $\sim(D \wedge \sim E)$	4. Law of *Modus Tollens* (2,3)
5. $\sim D \vee E$	5. De Morgan's Law
6. $\sim D \rightarrow C$ $\sim C$	6. Given
7. D	7. Law of *Modus Tollens*
8. E	8. Law of Disjunctive Inference (5,7)

42.

The plan: Prove that $DFBE$ is a parallelogram, then prove that $\overline{EB} \cong \overline{DF}$ because opposite sides of a parallelogram are congruent.

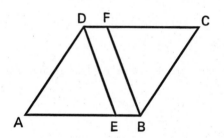

Statements	Reasons
1. *ABCD* is a parallelogram	1. Given
2. $\overline{DC}$ is parallel to $\overline{AB}$	2. Definition of a parallelogram
3. $\angle CFB \cong \angle ABF$	3. Alternate interior angles are congruent.
4. m $\angle ADC$ = m $\angle ABC$	4. Opposite angles of a parallelogram are equal in measure.
5. $\overline{ED}$ bisects $\angle ADC$; $\frac{2}{3}$ bisects $\angle ABC$	5. Given
6. m $\angle CDE = \frac{1}{2}$ m $\angle ADC$ m $\angle ABF = \frac{1}{2}$ m $\angle ABC$	6. Definition of angle bisector
7. m $\angle CDE$ = m $\angle ABF$	7. Halves of equal quantities are equal
8. m $\angle CDE$ = m $\angle CFB$	8. Transitive Property of Equality
9. $\overline{DE}$ is parallel to $\overline{FB}$	9. If two lines are cut by a transversal and corresponding angles are congruent, then the lines are parallel.
10. *DFBE* is a parallelogram	10. Definition of a parallelogram
11. $\overline{EB} \cong \overline{DF}$	11. Opposite sides of a parallelogram are congruent.

Tables of Natural Trigonometric Functions
(For use with Sequential Math – Course II Regents Examinations)

Angle	Sine	Cosine	Tangent	Angle	Sine	Cosine	Tangent
1°	.0175	.9998	.0175	46°	.7193	.6947	1.0355
2°	.0349	.9994	.0349	47°	.7314	.6820	1.0724
3°	.0523	.9986	.0524	48°	.7431	.6691	1.1106
4°	.0698	.9976	.0699	49°	.7547	.6561	1.1504
5°	.0872	.9962	.0875	50°	.7660	.6428	1.1918
6°	.1045	.9945	.1051	51°	.7771	.6293	1.2349
7°	.1219	.9925	.1228	52°	.7880	.6157	1.2799
8°	.1329	.9903	.1405	53°	.7986	.6018	1.3270
9°	.1564	.9877	.1584	54°	.8090	.5878	1.3764
10°	.1736	.9848	.1763	55°	.8192	.5736	1.4281
11°	.1908	.9816	.1944	56°	.8290	.5592	1.4826
12°	.2079	.9781	.2126	57°	.8387	.5446	1.5399
13°	.2250	.9744	.2309	58°	.8480	.5299	1.6003
14°	.2419	.9703	.2493	59°	.8572	.5150	1.6643
15°	.2588	.9659	.2679	60°	.8660	.5000	1.7321
16°	.2756	.9613	.2867	61°	.8746	.4848	1.8040
17°	.2924	.9563	.3057	62°	.8829	.4695	1.8807
18°	.3090	.9511	.3249	63°	.8910	.4540	1.9626
19°	.3256	.9455	.3443	64°	.8988	.4384	2.0503
20°	.3420	.9397	.3640	65°	.9063	.4226	2.1445
21°	.3584	.9336	.3839	66°	.9135	.4067	2.2460
22°	.3746	.9272	.4040	67°	.9205	.3907	2.3559
23°	.3907	.9205	.4245	68°	.9272	.3746	2.4751
24°	.4067	.9135	.4452	69°	.9336	.3584	2.6051
25°	.4226	.9063	.4663	70°	.9397	.3420	2.7475
26°	.4384	.8988	.4877	71°	.9455	.3256	2.9042
27°	.4540	.8910	.5059	72°	.9511	.3090	3.0777
28°	.4695	.8829	.5317	73°	.9563	.2924	3.2709
29°	.4848	.8746	.5543	74°	.9613	.2756	3.4874
30°	.5000	.8660	.5774	75°	.9659	.2588	3.7321
31°	.5150	.8572	.6009	76°	.9703	.2419	4.0108
32°	.5299	.8480	.6249	77°	.9744	.2250	4.3315
33°	.5446	.8387	.6494	78°	.9781	.2079	4.7046
34°	.5592	.8290	.6745	79°	.9816	.1908	5.1446
35°	.5736	.8192	.7002	80°	.9848	.1736	5.6713
36°	.5878	.8090	.7265	81°	.9877	.1564	6.3138
37°	.6018	.7986	.7536	82°	.9903	.1392	7.1154
38°	.6157	.7880	.7813	83°	.9925	.1219	8.1443
39°	.6293	.7771	.8098	84°	.9945	.1045	9.5144
40°	.6428	.7660	.8391	85°	.9962	.0872	11.4301
41°	.6561	.7547	.8693	86°	.9976	.0698	14.3007
42°	.6691	.7431	.9004	87°	.9986	.0523	19.0811
43°	.6820	.7314	.9325	88°	.9994	.0349	28.6363
44°	.6947	.7193	.9657	89°	.9998	.0175	57.2900
45°	.7071	.7071	1.0000	90°	1.0000	.0000	

Glossary

If a term you're looking for doesn't appear here, it may be defined in the Stuff You Should Know chapter. Otherwise, consult your math textbook.

A

abscissa   Another name for the x-coordinate of a point.

acute angle   An angle that measures less than 90°.

acute triangle   A triangle with three acute angles.

additive inverse   The opposite value of a number (which, when added to that number, yields zero). For example, the additive inverse of a is $-a$, because $a + (-a) = 0$.

alternate interior angles   Two interior angles that are on opposite sides of a transversal that intersects with two parallel lines.

altitude   A segment drawn from a vertex of a polygon (usually a triangle) that is perpendicular to the opposite side of that polygon.

associative property   Mathematical rule that states: $a + (b + c) = (a + b) + c$ and $a \times (b \times c) = (a \times b) \times c$.

axis of symmetry   The line that contains the vertex of a parabola and cuts the parabola exactly in half.

B

backsolving   The practice of plugging the answers provided on a multiple-choice test into the question to see which one works.

base angle   An angle that includes the base of a polygon (usually a triangle).

bisect   To cut exactly in half.

C

collinear   Located in the same line.

combinations   The number of ways you can choose a specific number of items in no particular order from a group.

commutative property Mathematical rule that states: $a + b = b + a$ and $a \times b = b \times a$.

complementary angles Two angles whose sum is 90°.

congruent The same size and shape (symbolized by $\cong$).

corresponding angles Two angles that appear in the same position when two lines are cut by a transversal.

D

denominator The bottom number of a fraction.

diameter The greatest distance within the circumference of a circle.

dilation The process under which the coordinates of each point in a figure are multiplied by a constant.

distributive property Mathematical property that states: $a(b + c) = ab + ac$.

E

equidistant The same distance away from a point or series of points.

equilateral triangle A triangle with three congruent sides and three angles that measure 60°.

exterior angle An angle that is formed when a side of a triangle is extended beyond the vertex of the triangle and is supplementary to its adjacent interior angle. (It also equals the sum of the triangle's other two non-adjacent interior angles.)

F

FOIL An acronym for First, Outer, Inner, Last, a process by which you multiply algebraic terms.

H

hypotenuse The longest side of a right triangle.

I

identity element   An element within an operation.

image   The result after a point or series of points undergoes a transformation.

intercept   The point at which a graph intersects one of the coordinate axes.

internal angle   An angle that lies in the interior of a triangle.

isosceles trapezoid   A trapezoid with two non-parallel sides that are congruent.

isosceles triangle   A triangle with at least two congruent sides.

L

leg   One of the two perpendicular sides of a right triangle.

locus   A set of points.

M

median of a triangle   A segment drawn from the vertex of a triangle to the midpoint of the opposite side of that triangle.

midpoint   The point equidistant from the endpoints of a line segment.

multiplicative inverse   The reciprocal of a number that is not equal to zero. For example, $\frac{1}{a}$ is the multiplicative inverse of a, because $a \times \frac{1}{a} = 1$.

N

negation   Asserting that a statement is false (A becomes $\sim A$).

negative reciprocals   Two numbers whose product is -1.

numerator   The top number in a fraction.

O

obtuse angle   An angle that measures greater than 90°.

obtuse triangle   A triangle that contains an obtuse angle.

operation   A process, such as addition or multiplication, by which numbers or variables are combined.

ordered pair   Two numbers written in a specific order (usually involving the coordinates of a point).

ordinate   Another name for the y-coordinate of a point.

origin   The point (0, 0) on the coordinate axes.

P

parallel lines   Lines within the same plane that have the same slope and will never intersect

parallelogram   A quadrilateral with two pairs of opposite sides that are parallel.

perfect square   A number or term with a square root that is rational.

perimeter   The sum of the lengths of all the sides of a polygon.

permutations   The number of ways in which a certain number of items can be displayed or arranged.

perpendicular lines   Two lines that intersect in a right angle.

plugging in   The process of replacing variables with numbers to turn an algebraic problem into an arithmetic problem.

process of elimination (POE)   Arriving at the right answer by eliminating all the other answer choices that you know are incorrect.

proportion   An equation you set up when the relationship between two pairs of numbers is the same.

Q

quadrilateral   A polygon with four sides.

R

radical sign   Another word for the root of a number (in this book, it means the square root and is denoted by the $\sqrt{\ }$ sign).

radicand   The number that appears beneath the radical sign.

radius   The distance from the center of a circle to the circumference of that circle.

rational number   A number that can be expressed as the quotient of two integers.

reflection   A transformation in which a point is "reflected" in a line, usually the x- or y-axis.

regular polygon   A polygon in which the sides and angles are congruent.

rhombus   A quadrilateral with four congruent sides.

right angle   An angle formed by two perpendicular lines that measures $90°$.

right triangle   A triangle that contains a right angle.

root   A number that makes an equation true. For example, 2 and –2 are the roots of the equation $x^2 - 4 = 0$.

Rule of 180   The sum of the three angles in a triangle is $180°$.

S

scalene triangle   A triangle with three unequal sides.

segment   A finite linear connection between two points.

similar triangles   Two triangles that have the same shape but not the same size (corresponding angles are congruent and corresponding sides are proportional).

SOHCAHTOA   Abbreviation for the relationships of the three main trigonometric ratios.

supplementary angles   Two angles whose sum is $180°$.

system of equations   Two or more equations involving the same variables.

T

T-chart   A list of coordinates of a particular graph.

translation   A transformation in which you add to or subtract from the coordinates of a point, thus mapping it onto its image, which is a specific distance away.

transversal   A line that cuts through two parallel lines, thus creating several pairs of congruent angles.

trapezoid   A quadrilateral with exactly two parallel sides.

turning point   The point at which the graph of a parabola changes direction (also known as a vertex).

V

vertex   (1) The point at which the graph of a parabola changes direction (also known as the turning point); (2) the point of a polygon; (3) the center point of an angle.

vertex angle   The angle in an isosceles triangle that is not equal to either of the other two angles.

vertical angles   Two opposite angles formed by two intersecting angles.

About the Author

Doug French graduated from the University of Virginia and has been working as a teacher, writer, editor, and course developer with The Princeton Review since 1991. He has taught classes for the PSAT, SAT, LSAT, GMAT, and GRE in the U.S., Europe, and Asia, and he has tutored math students in everything from fifth-grade arithmetic to BC calculus.

Doug also works as a freelance writer, draws cartoons, and does voice-overs. (He sounds a lot like that MovieFone guy.) His mom, however, has more talent in her little finger than he has in his little finger.

Free!

Did you know that The Microsoft Network gives you one free month?

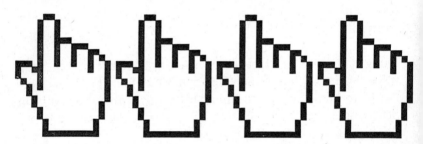

Call us at 1-800-FREE MSN. We'll send you a free CD to get you going.

Then, you can explore the World Wide Web for one month, free. Exchange e-mail with your family and friends. Play games, book airline tickets, handle finances, go car shopping, explore old hobbies and discover new ones. There's one big, useful online world out there. And for one month, it's a free world.

Call **1-800-FREE MSN,** Dept. 3197, for offer details or visit us at **www.msn.com.** Some restrictions apply.

Microsoft Where do you want to go today?®

The Microsoft Network

©1997 Microsoft Corporation. All rights reserved. Microsoft, MSN, and Where do you want to go today? are either registered trademarks or trademarks of Microsoft Corporation in the United States and/or other countries.

www.review.com

Expert Advice

Counselor-O-Matic

Pop Surveys

Paying for it

www.review.com

THE PRINCETON REVIEW

Getting In

Word du Jour

College Talk

Find-O-Rama College Search

www.review.com

MSN
The Microsoft Network
Includes FREE Offer

SAT Survival

Best Schools

www.review.com

CRACKING THE REGENTS!

Practice exams and test-cracking techniques

CRACKING THE REGENTS EXAMS: BIOLOGY
1998-99 • 0-375-75071-1 • $5.95

CRACKING THE REGENTS EXAMS: CHEMISTRY
1998-99 • 0-375-75072-X • $5.95

CRACKING THE REGENTS EXAMS: GLOBAL STUDIES
1998-99 • 0-375-75069-X • $5.95

CRACKING THE REGENTS EXAMS: SEQUENTIAL MATH I
1998-99 • 0-375-75064-9 • $5.95

CRACKING THE REGENTS EXAMS: SEQUENTIAL MATH II
1998-99 • 0-375-75065-7 • $5.95

CRACKING THE REGENTS EXAMS: SEQUENTIAL MATH III
1998-99 • 0-375-75066-5 • $5.95

CRACKING THE REGENTS EXAMS: SPANISH
1998-99 • 0-375-75067-3 • $5.95

CRACKING THE REGENTS EXAMS: U.S. HISTORY
1998-99 • 0-375-75068-1 • $6.95

CRACKING THE REGENTS EXAMS: COMP ENGLISH
1998-99 • 0-375-75083-5 • $5.95

CRACKING THE REGENTS EXAMS: EARTH SCIENCE
1998-99 • 0-375-75070-3 • $5.95

THE PRINCETON REVIEW

Available at your local bookstore, or order direct by
calling 1-800-733-3000, Monday-Friday, 8:30 am-5pm EST

Random House